ADVANCES IN PROTEIN CHEMISTRY AND STRUCTURAL BIOLOGY

EDITED BY

ALEXANDER McPHERSON
University of California, Irvine
Department of Molecular Biology and Biochemistry
USA

ELSEVIER

AMSTERDAM • BOSTON • HEIDELBERG • LONDON
NEW YORK • OXFORD • PARIS • SAN DIEGO
SAN FRANCISCO • SINGAPORE • SYDNEY • TOKYO
Academic Press is an imprint of Elsevier

Academic Press is an imprint of Elsevier
The Boulevard, Langford Lane, Kidlington, Oxford OX5 1GB, UK
30 Corporate Drive, Suite 400, Burlington, MA 01803, USA
525 B Street, Suite 1900, San Diego, CA 92101-4495, USA

First edition 2010

ISBN: 978-0-12-381278-0
ISSN: 1876-1623

For information on all Academic Press publications
visit our website at www.elsevierdirect.com

Printed and bound in USA
10 11 12 10 9 8 7 6 5 4 3 2 1

Contents

Genomic Studies and Computational Predictions of Nucleosome Positions and Formation Energies

Denis Tolkunov and Alexandre V. Morozov

Protein Homorepeats: Sequences, Structures, Evolution, and Functions

Julien Jorda and Andrey V. Kajava

Protein Aggregation: From Inclusion Bodies to Amyloid and Biomaterials

Anna Mitraki

Taking Charge of Proteins: From Neurodegeneration to Industrial Biotechnology

Bryan F. Shaw, Demetri T. Moustakas, Julian P. Whitelegge, and Kym F. Faull

Histone Acetylation, Acetyltransferases, and Ataxia—Alteration of Histone Acetylation and Chromatin Dynamics is Implicated in the Pathogenesis of Polyglutamine-Expansion Disorders

Shaun D. McCullough and Patrick A. Grant

GENOMIC STUDIES AND COMPUTATIONAL PREDICTIONS OF NUCLEOSOME POSITIONS AND FORMATION ENERGIES

By DENIS TOLKUNOV AND ALEXANDRE V. MOROZOV

Department of Physics & Astronomy and BioMaPS Institute for Quantitative Biology, Rutgers University, Piscataway, NJ, USA

Abstract

Chromatin is a complex of DNA, RNA, and proteins whose primary function is to package genomic DNA into the tight confines of a cell nucleus. A fundamental repeating unit of chromatin is the nucleosome, an octamer of histone proteins around which 147 base pairs of DNA are wound in almost two turns of a left-handed superhelix. Chromatin is a dynamic structure that exerts profound influence on regulation of gene expression and other cellular functions. These chromatin-directed processes are facilitated by optimizing nucleosome positions throughout the genome and by remodeling nucleosomes in response to various external and internal signals such as environmental perturbations. Here we discuss the large-scale maps of nucleosome positions made available through recent advances in parallel high-throughput sequencing and microarray technologies. We show that these maps reveal common features of nucleosome organization in eukaryotic genomes. We also survey the computational models designed to predict nucleosome formation scores or energies and

DOI: 10.1016/S1876-1623(10)79001-5

demonstrate how these predictions can be used to position multiple nucleosomes on the genome without steric overlap.

I. Introduction

DNA in eukaryotic nuclei is assembled into chromatin – a complex combination of DNA, RNA, and proteins that makes up chromosomes. The primary function of chromatin is to compact genomic DNA, which otherwise would not fit into the cell nucleus. However, since the early days of chromatin studies (Kornberg and Thomas, 1974) it has been recognized that chromatin's function goes well beyond DNA compaction; in particular, chromatin exerts a profound influence on gene regulation, replication, and recombination, and on DNA repair both by blocking access to DNA (Boeger et al., 2003) and by juxtaposing sites far apart on the linear sequence (Wallrath et al., 1994).

The building block of chromatin is the nucleosome core particle (Khorasanizadeh, 2004), a 147-base pair (bp) long DNA segment wrapped in ~1.65 superhelical turns around the surface of a histone octamer (Luger et al., 1997; Richmond and Davey, 2003) (Fig. 1A). On the first level of compaction, DNA is arranged into one-dimensional (1D) quasi-periodic nucleosomal arrays, which in turn fold into higher order chromatin fibers (Felsenfeld and Groudine, 2003). Chromatin fiber formation is stabilized in part by the linker histone H1. Neighboring nucleosomes (which consist of the nucleosome core particle and H1) are separated from each other by ~10–60 bp stretches of linker DNA, which means that ~70–90% of genomic DNA is wrapped in nucleosomes. The histone octamer is made of two copies of four highly conserved histone proteins: H2A, H2B, H3, and H4. Histones have unstructured tail domains that protrude from the surface of the histone octamer, providing sites for potential interactions with other proteins. Histone tails are targets of numerous post-translational covalent modifications such as acetylation, phosphorylation, methylation, ubiquitination, and ADP-ribosylation, and may also influence how nucleosome arrays fold into higher order chromatin structures (Strahl and Allis, 2000; Khorasanizadeh, 2004).

In this review we focus on the genome-wide predictions of nucleosome formation energies and positions obtained by analyzing DNA conformational properties and high-throughput nucleosome-mapping data. DNA

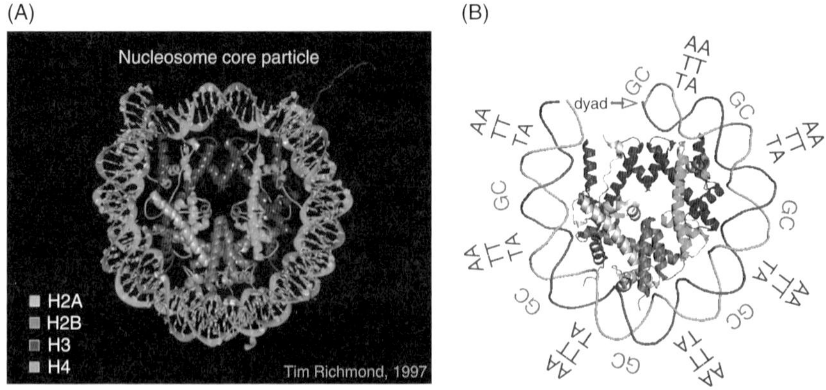

FIG. 1. (A) Crystal structure of the nucleosome core particle (Luger et al., 1997) (courtesy of Song Tan, Penn State University). The 147-bp long DNA is wrapped around the histone octamer in ~1.65 turns of a left-handed superhelix. The histone octamer consists of two copies each of histones H2A (yellow), H2B (red), H3 (blue), and H4 (green). (B) Bending of nucleosomal DNA is mediated by specific dinucleotides located at positions where DNA minor or major groove faces the histone octamer (reproduced with permission from Segal et al., 2006). Relative frequencies of A/T-rich dinucleotides tend to increase at positions where the minor groove faces the surface of the histone octamer, whereas relative frequencies of G/C-rich dinucleotides tend to increase where the minor groove faces away from and the major groove faces toward the histone octamer. (See color plate 1).

sequences differ greatly in their ability to form nucleosomes – *in vitro* studies show that the range of histone–DNA-binding affinities is at least a thousand-fold (Thastrom et al., 1999). Nucleosomal DNA is sharply bent to achieve tight wrapping around the histone octamer. This bending occurs at every 10–11 bp DNA helical repeat, where the minor groove of the DNA faces inward toward the histone octamer, and again at ~5 bp away, with opposite direction, when the minor groove faces outward (Fig. 1B). The bends of each direction are facilitated by specific dinucleotides – up to higher order effects, sequence-specific DNA bending is controlled by base stacking energies between neighboring base pairs.

It is reasonable to assume that *in vitro* nucleosome positions are determined purely by intrinsic sequence preferences and by steric exclusion between neighboring nucleosomes. *In vivo*, however, nucleosomes compete with non-histone DNA-binding factors for access to genomic DNA, which may result in overriding the intrinsic sequence preferences.

In addition, chromatin-remodeling enzymes play a role that needs to be quantified: in one scenario the role of such enzymes is purely catalytic, modifying the rate of assembly but not the final disposition of nucleosomes on DNA. In the other, ATP-dependent chromatin-remodeling enzymes actively reposition nucleosomes to control access to DNA, in analogy with motor proteins. The relative importance of intrinsic sequence preferences, chromatin-remodeling enzymes, competition with other factors, and formation of higher order structures for shaping and maintaining *in vivo* chromatin continues to be debated.

To address these questions, large-scale maps of nucleosome positions have been generated in recent years. Using microarray and massively parallel sequencing technologies, nucleosomes have been mapped genome-wide in *Saccharomyces cerevisiae* (both *in vivo* and *in vitro*), *Drosophila melanogaster, Caenorhabditis Elegans*, and *Homo sapiens*. These data were used to train bioinformatics models that attempt to predict nucleosome occupancy profiles and in particular discriminate between nucleosome-enriched and nucleosome-depleted regions using various sequence features (*e.g.* dinucleotide frequencies found in the alignment of experimentally mapped nucleosomal sequences (Segal et al., 2006)) as input. On the other hand, there are a number of models that do not rely on high-throughput data sets – rather, they employ a physics-based description of the nucleosome core particle to predict sequence-dependent elastic energies of bending nucleosomal DNA into a superhelix. The elastic energy is represented as a quadratic potential, using the empirical parameters obtained from DNA structural data, measurements of DNA mobility on a gel, and so on. The resulting energy profile is then used to predict nucleosome occupancies and positions, and compare them with experimental data.

In this review we focus on high-throughput nucleosome-positioning data sets and on the computational models created to explain them. The scope of this review does not allow us to discuss smaller scale nucleosome-positioning studies that had existed well before the first microarray-based nucleosome map was published in 2005. Although we have tried to review all the major data sets published between 2005 and 2009, rapid progress in parallel sequencing technology makes it certain that many more nucleosome-positioning maps will be available in the near future. The current data sets, however, have already been invaluable for understanding major features of nucleosome organization in eukaryotic genomes. They have also provided insight into how chromatin is shaped by genomic sequence features that dictate nucleosome positioning *in vitro*. In the modeling part of the review we focus on bioinformatics models and on physical models that compute sequence-dependent free energies of nucleosome formation.

This review is organized as follows: Section II.A covers nucleosome-positioning data and describes select observations drawn from it, focusing especially on stereotypical features of nucleosome organization in genic and intergenic regions. Section II.B gives a brief overview of the structural studies of the nucleosome core particle. Section III.A describes nucleosome models based on elastic-energy calculations. Section III.B summarizes the currently available bioinformatics models for chromatin structure prediction. Section III.C explains how statistical physics of 1D liquids can be applied to positioning multiple nucleosomes simultaneously without steric overlap. Section III.D describes a hidden Markov model (HMM) approach for inferring nucleosome occupancy from log-intensity microarray profiles. Finally, Section IV contains a brief summary of our main observations.

II. Experimental Studies of Chromatin Structure

A. *Genome-Wide Mapping of Nucleosome Positions*

1. *Nucleosome-positioning Studies in* S. cerevisiae

1.1. Microarray studies Given the current wealth of technologies for genome-wide mapping of nucleosome positions, it is amazing to note that these technologies date only from 2005. Prior to that, microarray resolution was simply too low (~1 kbp) to detect single nucleosome positions. Despite this limitation, early work showed a general depletion of nucleosomes from promoter regions (Bernstein et al., 2004; Lee et al., 2004). This important observation was refined in a pioneering 2005 study by Yuan et al. (2005) that employed a microarray to map nucleosome positions across 482 kb of the budding yeast genome, spanning almost the entire chromosome III and 223 additional regulatory regions. The microarray consisted of 50 bp oligonucleotide probes tiled every 20 bp across the genomic regions of interest. Nucleosomal DNA was digested with micrococcal nuclease (MNase; an endo–exonuclease that preferentially digests linker DNA) and labeled with green (Cy3) fluorescent dye. The sample was then mixed with total genomic DNA labeled with red (Cy5) fluorescent dye and hybridized to the microarray. The output of the experiment consisted of the log2 ratio of hybridization values for nucleosomal versus genomic DNA at each probe position across the genomic region of interest (see Fig. 2 for an overview of the method).

The log-intensity profiles served as input to an HMM (Section III.D), which was used to predict the probability of starting a nucleosome at each base pair and the corresponding nucleosome occupancies. (Nucleosome occupancy is defined as the probability that a given base pair is covered by any nucleosome and is therefore computed as the sum of the probability peaks for all nucleosomes that are close enough to overlap the base pair in question.) The authors confirmed the earlier low-resolution reports that intergenic DNA in yeast was nucleosome-depleted relative to coding DNA, and found nucleosome-depleted regions (NDRs) of ~150 bp in length immediately upstream of many annotated coding sequences (Fig. 3). Although the microarray resolution was insufficient for mapping individual nucleosomes with a bp-level precision, the authors were able to carry out a limited study of the sequence determinants of nucleosome positioning, and found that the nucleosome-free regions were enriched in poly-A and poly-T motifs. These motifs tend to occur in promoters, suggesting a causal role of poly(dA-dT) tracts in establishing NDRs.

Rapid progress in microarray technology allowed the first nucleosome map of the entire *S. cerevisiae* genome to be completed in 2007 (Lee et al., 2007). The Lee et al. study employed high-density Affymetrix tiling microarrays with 25-bp probes spaced every 4 bp across the yeast genome. Similarly to the earlier Yuan et al. (2005) study, genomic chromatin was cross-linked with formaldehyde and treated with MNase, resulting in preferential digestion of nucleosome-free linker sequences. Three independent samples of nucleosomal and total genomic DNA were hybridized to the microarray. The resulting log2-intensity traces were processed by the HMM to yield genome-wide nucleosome probabilities and occupancies. The authors used the Viterbi algorithm (Durbin et al., 1998) to identify nucleosome positions in the "top" (most likely) nucleosome configuration. Using a single configuration rather than a sum over all possible configurations allows placing nucleosomes uniquely but discards probabilistic information about alternative nucleosome positions

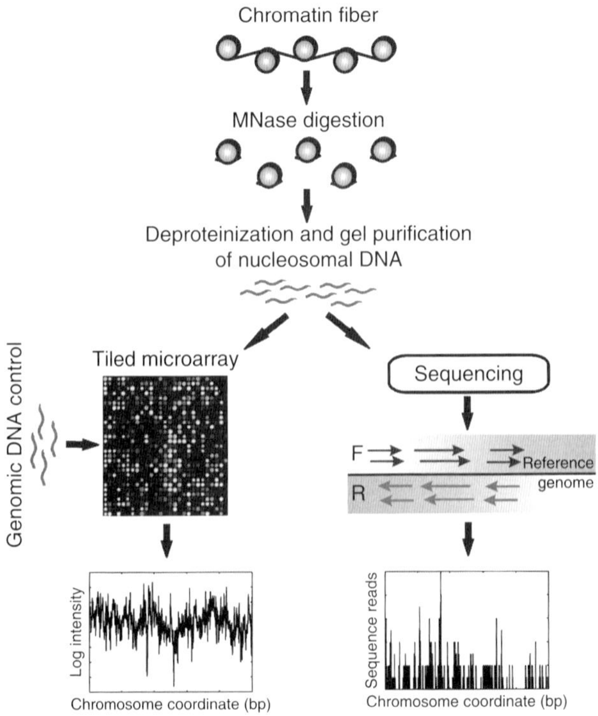

FIG. 2. Schematic representation of high-throughput nucleosome-positioning experiments. Chromatin is subjected to MNase digestion, the resulting DNA is purified, and mononucleosomal DNA is isolated on a gel. Optionally, histones are cross-linked on DNA prior to the MNase digestion and immunoprecipitated, after which the cross-linking is reversed. In a chip-based experiment (left panel), mononucleosomal DNA is hybridized to a microarray together with a control sample prepared by digesting genomic nucleosome-free DNA with MNase. The log ratio of nucleosomal DNA intensity to genomic DNA intensity is plotted as a function of each probe's starting position on the genome. Higher-than-average log-intensity values correspond to nucleosome-covered regions, while lower-than-average log-intensity values correspond to nucleosome-depleted regions. Microarray probes are tiled across the entire genome or genomic regions of interest. In a high-throughput parallel sequencing experiment, mononucleosomal (and in some cases control) DNA is sequenced directly (right panel). High-throughput sequencing yields large collections of reads typically shorter than the DNA lengths in the input sample. The reads are mapped onto both strands of the reference genome (often with several mismatches allowed) and combined into a single sequence-read profile by assuming that each nucleosome core particle has a fixed length of 147 bp. With this assumption, sequence reads from the reverse strand are remapped onto the forward strand by subtracting 147 bp from their start coordinates. The resulting sequence-read profile contains information about the number of reads assigned to every genomic position. (See color plate 2).

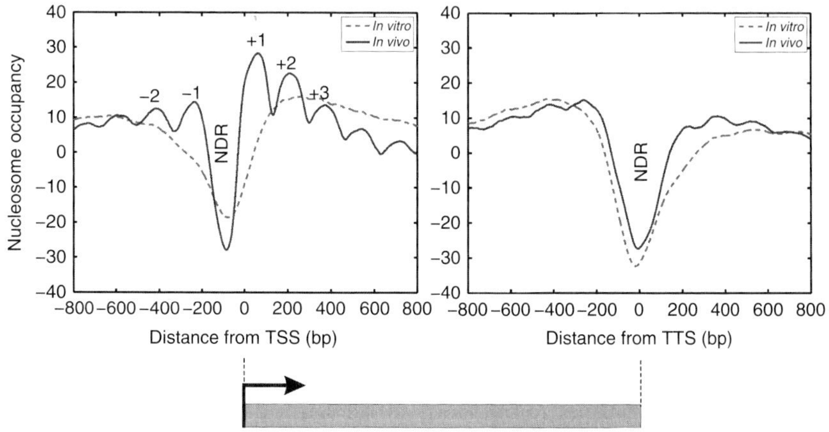

FIG. 3. Nucleosome occupancy in the vicinity of transcription start sites (TSSs) and transcription termination sites (TTSs) in *S. cerevisiae*. The unnormalized occupancy is defined as the number of nucleosomes covering a given base pair. Solid lines: *in vivo* occupancy (YPD medium, average over four replicates without cross-linking), dashed lines: *in vitro* occupancy (average over two replicates). *In vivo* and *in vitro* nucleosome positions were mapped by Kaplan et al. using high-throughput sequencing (Kaplan et al., 2009). Transcript coordinates are from Nagalakshmi et al. (2008). The genome-wide average of the nucleosome occupancy is subtracted from the plots.

contained in the full HMM approach (Section III.D). The authors confirmed existence of NDRs on a genome-wide scale and observed that nucleosome occupancy profiles correlated with transcript abundance and transcription rate. In addition, functionally related genes were clustered on the basis of nucleosome occupancy patterns observed at their promoters.

1.2. High-throughput sequencing studies By 2007 massively parallel sequencing technologies had matured to a point where it became possible to sequence hundreds of thousands of nucleosomal DNA molecules directly instead of hybridizing them to a microarray (see Fig. 2 for an overview of the method). The necessary technology, first developed by 454 Life Sciences (http://www.454.com), was capable of sequencing ~100 bp molecules, comparable to the 147-bp length of the nucleosome core particle. Thus it was natural to apply high-throughput sequencing to nucleosomal DNA. The first such study, carried out by Frank Pugh and coworkers, focused on *S. cerevisiae* nucleosomes with the histone variant H2A.Z (Albert et al., 2007). H2A.Z is conserved by evolution and is believed to be involved in transcriptional regulation, antisilencing, silencing, and genome stability (Draker and Cheung, 2009).

The Albert et al. study mapped 322,000 H2A.Z nucleosomes that had been fixed on DNA by formaldehyde cross-linking, immunoprecipitated, MNase-digested, and gel-purified. The most important difference from the microarray experiments was that instead of hybridizing nucleosomal DNA samples to the array (after further fragmenting the ~150-bp sequences with nuclease into ~50-bp pieces to make hybridization with relatively short probes more efficient), DNA molecules were sequenced directly and mapped onto the reference genome, providing a starting coordinate for each sequence read. This procedure resulted in a sequence read profile in which a non-negative number of reads was assigned to each genomic base pair (Fig. 2). Unlike microarray hybridization approaches, sequence read profiles can be constructed separately for both strands, and either compared with each other (Fig. 4) or combined.

The authors had to assume that the nucleosome position was given precisely by the coordinates of the sequence read, neglecting potential errors caused by imperfect MNase digestion. They also had to extend the mapped sequence reads to cover the 147-bp length of the nucleosome core particle (This procedure is common to all short-read sequencing studies.) Furthermore, because only the H2A.Z-containing nucleosomes were detected, the question of bulk nucleosome positioning remained open.

This question was addressed in a 2008 study by the same group (Mavrich et al., 2008a), which used immunoprecipitation with antibodies against tagged histones H3 and H4 to map 1,206,057 bulk nucleosomes from the yeast genome. The use of the 454 sequencing and the construction of the sequence read profile were the same as in the earlier H2A.Z study. These sequencing studies helped establish a "canonical" picture of nucleosome organization in which well-positioned –1 and +1 nucleosomes bracket an NDR upstream of *S. cerevisiae* genes (Fig. 3).

Mavrich et al. argued that positioning of bulk nucleosomes is largely a consequence of steric exclusion: +1 and to a certain extent –1 nucleosomes form barriers against which the other nucleosomes are "phased." In this scenario, sequence specificity would be important only for a small fraction of positioned nucleosomes. This picture is consistent with the

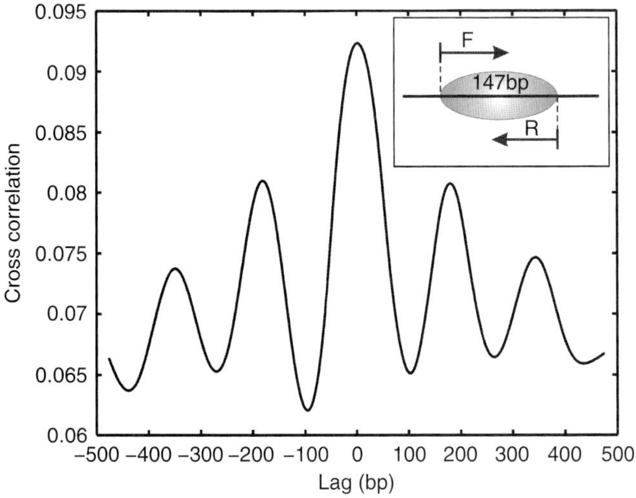

FIG. 4. Correlation function between the sequence reads mapped onto the Watson and Crick strands, respectively. High-throughput sequencing data is for *in vivo* nucleosomes in *S. cerevisiae* (Mavrich et al., 2008a). The observed maximum at zero lag corresponds to sequence reads on the opposite strands separated by 147 bp, thus demarcating the same nucleosome (see inset). Maxima at non-zero lag show relative positions of neighboring nucleosomes. The correlation function was smoothed using a 50 bp window average.

observation that nucleosomal dinucleotide patterns are more pronounced in the −1 and +1 nucleosomes than in the bulk ones (Mavrich et al., 2008a). The observed patterns are non-periodic and consist of a gradient of the TA/AA/TT dinucleotide counts at positions covered by −1 and +1 nucleosomes. The gradient is absent in bulk nucleosomes and its direction corresponds to a decrease in the TA/AA/TT counts with distance to the NDR (*i.e.* the counts increase in the 3′ direction for the −1 nucleosome and in the 5′ direction for the +1 nucleosome).

In addition to the 5′ NDR, a novel 3′ NDR (Fig. 3) that coincides with the transcription termination site (TTS) was discovered by the authors and it was argued that it may be implicated in transcription termination, antisense initiation, and gene looping. They concluded that the terminal nucleosomes (3′ nucleosomes immediately upstream of the TTS) may be partially positioned by sequence, including nearby cleavage and polyadenylation sites (AATAAA and related sequences).

Another *in vivo* map of nucleosome positions in *S. cerevisiae* was published by Eran Segal and coworkers in 2008 (Field et al., 2008). The authors used the 454 parallel pyrosequencing technology to sequence 503,264 yeast nucleosomes. The nucleosomes were mapped to the yeast genome by BLAST with a 95% sequence identity cutoff. The authors also required that sequence reads map to a unique location, have a length between 127 and 177 bp, and do not overlap with the ribosomal RNA locus. The resulting 378,686 nucleosomes were retained for further analysis. The 454 pyrosequencing technology used in this study was capable of creating ~200-bp reads (longer than the ~100-bp reads obtained by Frank Pugh and colleagues (Albert et al., 2007; Mavrich et al., 2008a)) and thus mononucleosome fragments were sequenced in full. However, even though both ends of the read mononucleosome were known, a wide distribution of fragment lengths made it impossible to predict individual nucleosome positions with a bp-level precision. Furthermore, there was no immunoprecipitation step, and the sequence read coverage was approximately a quarter of the 454 data set created by Frank Pugh and colleagues (Mavrich et al., 2008a).

The main focus of the Field et al. paper is on the nucleosome-positioning signals, which the authors captured with a model based on: a) relative dinucleotide frequencies at 127 central positions in the alignment of all sequence reads around their center (Segal et al., 2006); b) ratios of 5-mer frequencies in the linkers (defined as contiguous regions of 50–500 bp not covered by any nucleosome) to those in the nucleosome-bound sequences. The model is described in more detail in Section III.B.1; here we simply note that AAAAA/TTTTT was found to be the 5-mer with the strongest enrichment in linkers. The authors attribute a significant nucleosome-positioning role to this and other A/T-rich "boundary zone" elements that tend to be underrepresented in nucleosomes but overrepresented immediately outside of the nucleosome cores in the linker regions.

The authors also argue that nucleosome depletion over A/T-rich "boundary elements" is unlikely to be an MNase artifact, on the basis of the low rank-order correlation observed between word frequencies across MNase cut sites and relative word frequencies in linkers versus nucleosomes. In particular, AAAAAA is ranked 1782nd as an MNase cleavage site and 1st for enrichment in linker regions. The authors also find that their data yields an average nucleosome occupancy profile with respect to the transcription start site (TSS) that is in a broad agreement with the earlier microarray study by Lee et al. (2007): for most genes, there is a prominent NDR flanked by oscillations in the nucleosome occupancy that are usually interpreted as being a consequence of steric exclusion.

Except for several studies focused on changes in chromatin structure with respect to environmental or genetic perturbations (cf. Section II.A.1.3), prior to 2009 all nucleosome maps had come from *in vivo* chromatin of yeast cells grown in rich YPD medium. However, in 2009 Eran Segal and coworkers employed the short-read Solexa/Illumina sequencing (http://www.illumina.com/sequencing) to compare the *in vitro* and *in vivo* nucleosome positions and to study how chromatin structure changes under different growth conditions (Kaplan et al., 2009). The *in vitro* map is especially important because nucleosome locations should be dictated purely by steric exclusion and intrinsic sequence preferences. For *in vivo* maps, yeast cells were grown in YPD medium as well as YP media supplemented with 2.0% galactose or 2.8% ethanol instead of glucose. For each medium, nucleosome DNA samples were prepared both with and without formaldehyde cross-linking and sequenced by synthesis using the Solexa/Illumina technique. For the *in vitro* map, yeast genomic DNA was purified and mixed with histone octamers from chicken erythrocytes. Nucleosomes were reconstituted by salt gradient dialysis (Thastrom et al., 2004) at a lower histone octamer concentration than that observed *in vivo* (40 μg histone octamer per 100 μg DNA). The lower histone concentration was necessary as reconstitutions at higher *in vivo* stoichiometry resulted in insoluble chromatin, which was inaccessible to MNase.

Comparison between nucleosome positions from *in vitro* and *in vivo* experiments revealed striking overall similarity, leading the authors to conclude that: a) *in vitro* and *in vivo* nucleosome maps are highly similar; and b) chromatin structure is largely invariant with respect to different growth conditions. The authors also concluded that nucleosome positions are largely encoded by intrinsic DNA sequence signals, because a purely sequence-dependent model fit on the *in vitro* data was able to predict *in vivo* nucleosome locations with reasonably high accuracy. The model is essentially identical to the earlier one from Field et al. (2008), and is described in more detail in Section III.B.1. Similarly to this previous study, the authors found using both *in vitro* and *in vivo* maps that

5-mers with the lowest average nucleosome occupancy were AAAAA and ATATA. In addition to the nucleosome-excluding and nucleosome-favoring distributions of 5-mers, a 10 to 11-bp periodic dinucleotide signal caused by DNA bending (as discussed in the Section I), with AA/AT/TA/TT frequencies out of phase with CC/CG/GC/GG frequencies (Fig. 5), was also described by the authors. These periodic oscillations have been observed in both *in vitro* and *in vivo* nucleosome-positioning sequences (NPSs) (Albert et al., 2007; Field et al., 2008; Mavrich et al., 2008a; Kaplan et al., 2009).

Finally, a nucleosome-depleted region is present *in vitro* at both the TSS and the translation end (which was chosen because the authors believe TTSs to be poorly annotated in yeast). Interestingly, though there is little difference between *in vivo* and *in vitro* nucleosome depletion at 3′ ends of genes, the 5′ NDR is much shallower for *in vitro* chromatin

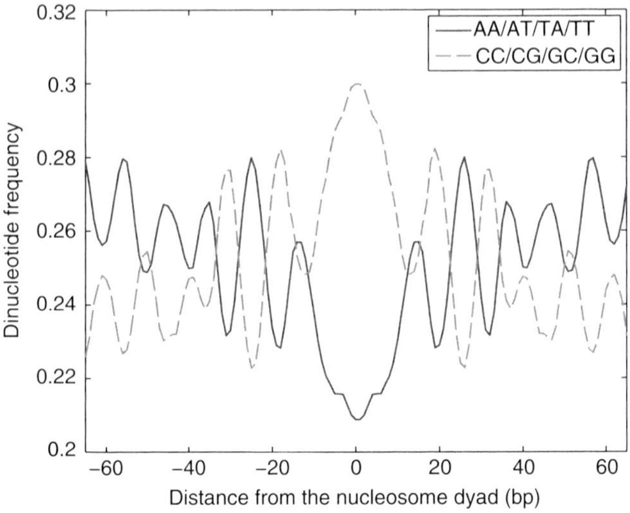

Fig. 5. Periodic dinucleotide frequencies observed in a high-throughput data set of nucleosome-positioning sequences in yeast (Kaplan et al., 2009). The 147-bp long *in vitro* nucleosome-positioning sequences defined by five or more sequence reads were aligned and the relative frequencies of AA/AT/TA/TT and CC/CG/GC/GG dinucleotides were computed at each position in the nucleosomal site. The frequencies were divided by genome-wide propensities for each group of dinucleotides.

and there are no characteristic oscillations in the flanking regions (Fig. 3). If these oscillations were indeed induced by steric exclusion, their absence from *in vitro* chromatin indicates that nucleosomes are not positioned as precisely, and suggests that intrinsic sequence signals are not the only contribution to the *in vivo* anchoring of nucleosomal arrays.

Another study that focused on mapping nucleosomes assembled *in vitro* on genomic DNA was carried out by Kevin Struhl and coworkers (Zhang et al., 2009). The authors purified both *S. cerevisiae* and *Escherichia coli* genomic DNA and assembled it into chromatin either by salt dialysis with *D. melanogaster* histones or by using an *in vitro* system containing recombinant *D. melanogaster* proteins NAP-1 (nucleosome assembly protein 1) and ACF, an ATP-dependent chromatin assembly factor known to produce arrays of regularly spaced nucleosomes (Ito et al., 1997). *In vitro* chromatin was subsequently digested with MNase to mononucleosome core particles, and the resulting DNA was purified and sequenced on a Solexa/Illumina Genome Analyzer, yielding 1–3 million uniquely mapped sequence reads for each input sample. As a control, the authors sonicated a mixture of yeast and *E. coli* DNA, followed by isolating and sequencing fragments of mononucleosomal size.

The authors discovered that nucleosomes strongly prefer yeast DNA to *E. coli* DNA, indicating that yeast genome evolved to facilitate nucleosome formation. As in the Kaplan et al. (2009) study, they found that many regions around TSSs and TTSs intrinsically disfavor nucleosome formation and that nucleosomes positioned *in vitro* by salt dialysis exhibit prominent periodic distributions of AA/TT/AT dinucleotides. In contrast, nucleosomes positioned by ACF had fewer NDRs and significantly less prominent periodicity of dinucleotide frequencies, showing that ACF is capable of overriding intrinsic sequence specificities of nucleosome core particles.

The main conclusion reached by Zhang et al. (2009) is that intrinsic histone–DNA interactions are not a major determinant of *in vivo* nucleosome positions. This is in contrast to the Kaplan et al. (2009) study, which argues that *in vivo* and *in vitro* nucleosome occupancy profiles are highly similar and that the latter can be explained with a model based purely on DNA sequence features. Zhang et al. focus instead on the fact that the *in vivo* pattern of statistical nucleosome positioning around 5′ NDRs is not observed *in vitro* and thus cannot be determined by intrinsic sequence preferences alone (Fig. 3). They argue that the strong positioning of the +1 nucleosome is linked to the process of transcriptional initiation, and propose that, although *in vitro* 5′ NDRs may facilitate assembly of the pre-initiation complex (PIC), an early step in the transcription process (probably preceding extensive elongation) is a primary determinant of the +1 nucleosome positioning. They hypothesize that some component of the transcriptional initiation machinery interacts with a nucleosome-remodeling complex and/or histones to position the +1 nucleosome. Once in place, the +1 nucleosome positions the +2, +3,... nucleosomes by steric exclusion. This view is supported by the observation that *in vivo* +2, +3,... nucleosomes are much better positioned than their –2, –3,... counterparts, even though the intrinsic positioning effect of the NDR should be the same on both sides.

1.3. Physiological and genetic perturbations of* S. cerevisiae *chromatin There are several studies of nucleosome positions in yeast whose main focus is not on intrinsic sequence preferences nor on the nucleosome organization with respect to various genomic features, but rather on how chromatin responds either to environmental perturbations or to deleting genes implicated in chromatin remodeling and maintenance.

The first high-throughput study of this kind was carried out in 2007 by Toshio Tsukiyama and coworkers (Whitehouse et al., 2007). The authors investigated the role of the ATP-dependent chromatin-remodeling complex Isw2 in controlling chromatin structure across the yeast genome. They sought to discover Isw2 targets genome-wide by identifying differences in nucleosome positions between wild-type and Δ*isw2* mutant strains. To this end, chromatin was cross-linked by formaldehyde, digested with MNase and exonuclease III, and purified to mononucleosomes using gel electrophoresis. Mononucleosomal DNA from both strains was separately hybridized to high-resolution Affymetrix tiling microarrays with ~5 bp probe spacing. Based on the difference in hybridization intensity between wild-type and mutant strains, the authors identified >1,000 regions, typically ~600 bp in length, where chromatin structure was disrupted in the Δ*isw2* mutant (these changes affected ~12% of yeast promoters). The authors also used chromatin immunoprecipitation (ChIP) to determine whether Isw2 was present at the loci whose nucleosome positions changed between strains.

The authors concluded that Isw2 functions by moving nucleosomes toward intergenic regions, where many important regulatory sequences are located. By doing so, it overrides intrinsic sequence-based nucleosome-positioning signals, as evidenced by the fact that poly(dA-dT) tracts are located within nucleosome +1 at many Isw2 targets (defined as promoters that exhibited change in the Δ*isw2* deletion strain). Loss of Isw2 would thus allow nucleosomes to relocate to their inherently preferred sites, lowering the total free energy of the system. The ability of Isw2 and other chromatin-remodeling enzymes to actively reposition nucleosomes demonstrates that intrinsic nucleosome-positioning preferences may be disrupted in living cells.

The Whitehouse et al. (2007) Isw2 study was followed in 2009 by a study by Hartley and Madhani (2009), which focused on how nucleosome positioning was affected by degrading proteins believed to be essential for maintaining chromatin structure: Myb family proteins Abf1 and Reb1, and the catalytic subunit of the RSC-remodeling complex, Sth1. As Reb1, Abf1, and Sth1 are all essential proteins, the authors had to use conditional alleles rather than gene deletion strains. Specifically, they used the temperature-sensitive degron system to engineer yeast strains in which protein degradation could be controlled via the N-end rule pathway (Dohmen and Varshavsky, 2005). Nucleosomal DNA samples were collected from wild-type and "degron" strains and interrogated using a low-resolution microarray designed to cover yeast chromosome III at 20 bp tiling steps (Yuan et al., 2005). The authors found that though the effect of Reb1 and Abf1 depletion was minimal, affecting ~10% of chromosome III promoters, depleting Sth1 affected the majority (~55%) of promoters. The affected genes displayed shrinking of the NDR accompanied by the movement of flanking nucleosomes. Although NDR was reduced in width, it was not eliminated, and the authors hypothesized that it is maintained by intrinsic sequence preferences. In support of this hypothesis, nucleosome positions were better predicted by the intrinsic "nucleosome positioning signature" (a first-generation nucleosome-positioning model by Ioshikes et al. (2006), cf. Section III.B.2) in the absence of Sth1.

It is also of interest to know how chromatin structure in yeast cells responds to physiological perturbations such as heat shock that are usually accompanied by massive transcriptional changes. The first study of this kind was carried out in 2008 by Vishwanath R. Iyer and colleagues (Shivaswamy et al., 2008). The authors subjected yeast cells grown in a rich medium to a 15-min period of heat shock. At the end of the 15-min period, control and heat-shocked cells were treated with formaldehyde.

Mononucleosomal DNA was isolated by means of a standard protocol that involves MNase digestions and gel purification, and sequenced using Solexa/Illumina short-read technology. In this way, the authors generated a differential map of nucleosome positions, which consisted of 514,803 and 1,036,704 uniquely mapped reads for the normal and heat-shock growth conditions, respectively.

As in the work by Frank Pugh and colleagues, which was published at approximately the same time (Mavrich et al., 2008a), the authors find both 5′ and 3′ NDRs, with a well-positioned nucleosome at the 3′ end of the coding region. Thus, yeast genes are demarcated by NDRs at each end of the transcribed region. Nucleosomes located next to the NDRs are well positioned, at least in part because NDRs act as barriers against which the genic nucleosomes are "phased."

The authors also addressed the question of whether positioning of terminal or bulk nucleosomes over transcribed regions could be attributed to intrinsic sequence signals. They discovered that the distribution of AA/TT dinucleotide frequencies is 10–11 bp periodic in +1 nucleosomes. Furthermore, this periodic profile is also observed in +2, +3,... nucleosomes, showing that their positioning may be maintained through sequence signals in addition to steric constraints.

Surprisingly, the majority of nucleosomes did not change positions upon transcriptional perturbation caused by heat shock, either in promoters or in the coding regions. At some promoters, remodeling events were observed that could be classified into eviction, appearance, or repositioning of one or two nucleosomes. However, there were no simple rules that controlled nucleosome remodeling at induced and repressed promoters. Thus, although gene activation was associated on average with nucleosome eviction and gene repression with nucleosome appearance, there were cases in which strongly positioned nucleosomes appeared at induced promoters. Furthermore, many nucleosome-remodeling events occurred at promoters that did not experience any significant transcriptional change.

Many of these findings were corroborated in a 2009 study in which global nucleosome positioning was examined before and after global transcriptional restructuring caused by adding glucose to yeast cells grown in glycerol, a poor carbon source (Zawadzki et al., 2009). This nutrient upshift creates significant changes in the gene expression of more than half of all yeast genes (Zaman et al., 2009). The authors isolated mononucleosomal DNA using standard MNase digestion methods at three time points: immediately before, 20 min after, and 60 min after adding glucose to yeast cells. Nucleosomal and total genomic DNA were hybridized to Affymetrix microarrays tiled each 4 bp across the yeast genome. The authors also developed an HMM approach to processing log-intensity profiles (cf. Section III.D), which they used to predict nucleosome occupancies across the yeast genome.

For most genes changes in expression were not associated with nucleosome addition, removal, or repositioning within their promoters, although for genes containing TATA boxes the correlation between change in gene expression and change in nucleosome occupancy was somewhat higher (0.48 for TATA genes vs. 0.34 for all genes). The promoters of only 10% of all genes gained or lost nucleosomes despite the fact that >50% of all genes exhibited a change in mRNA levels of twofold or more (Zaman et al., 2009). Thus it appears that *in vivo* interactions of transcription factor (TF)-binding sites with their cognate factors are largely dictated by pre-positioned nucleosomes and that regulation of gene expression through these sites is mediated by changes in local TF concentration rather than nucleosome addition or removal. The unaltered promoter nucleosome structure for most glucose-regulated genes implies the existence of constitutively accessible binding sites for the factors that control expression of these genes. This is consistent with the notion of "pre-set" chromatin, which plays a largely instructive role in regulating gene expression (Morse, 2007).

2. *Nucleosome-positioning Studies in Higher Eukaryotes*

2.1. **Caenorhabditis elegans** High-throughput sequencing and microarray methods pioneered with *S. cerevisiae* were soon applied to other eukaryotes. In two recent papers Andrew Z. Fire and coworkers mapped nucleosome positions in another model organism *C. elegans* (Johnson et al., 2006; Valouev et al., 2008). The first of these studies presented a collection of 284,091 nucleosome cores sequenced with the 454 pyrosequencing technology. The nucleosomes came from a mixed-stage population of *C. elegans*. As in yeast, nucleosome sequence reads were mapped onto the reference genome, with ~60% of all reads assigned unambiguously to genomic loci. The resulting map had the coverage of one nucleosome per 300–400 bp of genomic DNA. Analysis of dinucleotide distributions revealed a pronounced periodicity in AA and TT frequencies that extended across the nucleosome core.

Another observation concerned the effect of the MNase sequence specificity on nucleosome positioning: in agreement with the earlier study of MNase sequence preferences (Wingert and Von Hippel, 1968), the authors observed preferential cleavage at A/T-rich target sites, with G/C residues considerably underrepresented at both positions flanking the cleavage site. However, there was no corresponding A/T enrichment around position 147 (where the end of the nucleosome core particle would have been if it were mapped with a bp-level precision). The authors interpret this lack of symmetry as evidence that MNase sequence specificity influences the choice of cleavage sites but does not lead to nucleosome repositioning. Indeed, if nucleosomes were actively repositioned by interactions with MNase, both ends of the core particle would have been marked by A/T-rich flanking regions. As mentioned above, symmetric flanking regions were not observed in the data, even though this may be simply due to inaccuracies in locating the other end of the nucleosome.

This study was extended in 2008 by the same group using a massively parallel technique of sequencing by oligonucleotide ligation and detection (SOLiD by Applied Biosystems: solid.appliedbiosystems.com) (Valouev et al., 2008). Parallel sequencing yielded more than 44 million uniquely mapped nucleosome cores from a mixed-stage population of *C. elegans*. SOLiD sequencing platform produces 50-bp reads – shorter than the 147-bp length of the nucleosome core particle. As a result (and similarly to other sequencing studies), the position of the nucleosome dyad had to be inferred by adding 73 bp to the starting base pair for reads mapped onto the Watson strand and subtracting 73 bp from the starting base pair for reads mapped onto the Crick strand.

The authors found that the absolute nucleosome positions varied substantially, possibly reflecting a lack of universal sequence-dictated positioning across *C. elegans* cell types. Nonetheless, nucleosomes tended to be arranged in repeated arraylike structures, presumably due to steric constraints. Sequence analysis of nucleosome cores showed an oscillating ~10 bp periodicity for AA/TT with an out-of-phase ~10 bp periodicity for GC. Longer words (up to 6-mers) were also distributed non-randomly, with a pronounced enrichment of A/T nucleotides around sequence read starts, which the authors again ascribed to MNase sequence specificity.

2.2. **Drosophila melanogaster** In 2008 Frank Pugh and co-authors used the 454 pyrosequencing to map 652,738 H2A.Z-containing nucleosomes to 207,025 locations in the *D. melanogaster* genome (Mavrich et al., 2008b). Similarly to the yeast studies from the same group (Albert et al., 2007; Mavrich et al., 2008a), *Drosophila* embryos were treated with formaldehyde, H2A.Z-containing nucleosome core particles were immunopurified, and nucleosomal DNA was sequenced. Because *Drosophila* embryos consist of a wide variety of cell types, the nucleosome map is an average over cells with potentially very different gene expression profiles. Nonetheless, the nucleosome organization showed generic features that transcended the differences in cell types. The most prominent of these was the nucleosome-depleted region upstream of the TSS. There were two essential differences from the earlier study of H2A.Z nucleosomes in yeast (Albert et al., 2007): the absence of a well-positioned −1 nucleosome and longer linker lengths in the fly, manifested as larger distances between consecutive nucleosomal peaks downstream of the +1 nucleosome. In addition, the genic (+1, +2,...) array of nucleosomal peaks started ~75 bp further downstream from the equivalent position in *S. cerevisiae*, with potentially important implications in how the TSS is presented to RNA polymerase II (Pol II): in *S. cerevisiae* the TSS resides within the nucleosome border, whereas in *D. melanogaster* the TSS tends to be nucleosome-free. This is consistent with the hypothesis that gene regulation occurs predominantly at the level of transcript initiation in *S. cerevisiae*, whereas in *D. melanogaster* transcript elongation may play a more important role.

As in yeast, the 3′ ends of fly genes tend to be nucleosome-depleted. H2A.Z NPSs exhibit periodic, out-of-phase distributions of A/T-rich and G/C-rich dinucleotides. Finally, there is a correlation between AA/TT and CC/GG content and nucleosome positioning: nucleosome-covered positions tend to be G/C-rich, whereas 5′ and 3′ NDRs are enriched to some extent in A/T nucleotides, including poly(dA-dT) TTSs at the 3′ end of *Drosophila* genes. The same study produced a lower resolution map of bulk (both H2A- and H2A.Z-containing) nucleosomes by digesting chromatin with MNase and hybridizing DNA samples to Affymetrix *Drosophila* tiling arrays with an average probe spacing of 36 bp (there was no immunoprecipitation step in this assay). The same nucleosome-positioning pattern was found with respect to the TSS, with the exception of a distinct −1 peak, which was not present in the H2A.Z map. Thus in both yeast and fly

–1 nucleosomes are well-positioned but in fly they tend to be H2A.Z-free, resulting in the absence of a prominent –1 peak in H2A.Z maps.

2.3. **Homo sapiens** The first study of nucleosome positioning on human genome was carried out in 2007 using high-resolution microarrays. Using MNase digestion, Ozsolak et al. (2007) isolated mononucleosomal DNA from five types of human cells: primary fibroblasts (IMR90), primary melanocytes (PM), mammary epithelial cells (MEC), melanoma (A375, MALME), and breast cancer cell lines (T47D, MCF7). Nucleosome-free genomic DNA from the same cell line (digested to a similar size distribution) was used as control. Nucleosomal and genomic DNA samples were labeled by different fluorescent dyes (Cy5 and Cy3, respectively) and hybridized to microarrays containing 50-bp probes. The probes were tiled in 10-bp steps and spanned 1.5-kb repeat-masked promoter regions of 3,692 genes, including 1,346 genes in the Affymetrix Human Cancer G110 Array and 2,346 randomly selected genes.

Because the log-intensity profiles came in 1.5-kb fragments, the HMM algorithm could not be used for predicting nucleosome positions and occupancies (HMMs require contiguous input data (Durbin et al., 1998)). The authors chose to use wavelet-based de-noising instead, followed by an edge-detection algorithm. Because log-intensity microarray profiles measure nucleosome occupancies, application of edge-detection techniques amounts to predicting nucleosome positions from occupancy data.

In order to examine whether NDRs existed in human cells, Ozsolak et al. compared nucleosome organization in expressed and unexpressed promoters from A375, IMR90, and MALME cell lines. On average, NDRs of expressed genes were much more pronounced. Some of the unexpressed genes also had NDRs, which the authors attributed to the fact that chromatin structure was pre-modified, making those genes poised for rapid expression. This hypothesis was supported by an observation that NDRs of unexpressed genes were likely to have transcription PICs pre-assembled at their promoters. Conversely, unexpressed genes without PICs had no nucleosome depletion around the TSS.

Finally, the authors looked for short sequence motifs preferentially enriched or depleted in nucleosome-covered regions. They found that

TATAAA, TATATA, GCGCGC, and AAAAAA motifs were enriched in nucleosome linkers, whereas TTCGA and CTGCTG motifs were enriched in nucleosome cores. The authors argued that since none of the linker-enriched motifs corresponded to the previously published MNase recognition sequences (Horz and Altenburger, 1981), MNase sequence specificity did not exert a significant influence on detected nucleosome positions, nor did it bias which subset of nucleosomes was detected. This conclusion is similar to that reached in yeast by Field et al. (2008), but is at variance with the *C. elegans* studies (Johnson et al., 2006; Valouev et al., 2008).

Another nucleosome-positioning experiment was published in 2007 by Robert E. Kingston and coworkers (Dennis et al., 2007). The authors employed two complementary approaches for mapping nucleosomes in human genome: a tiling microarray and a capillary electrophoresis-based sequencing. The microarray was custom-designed by NimbleGen Systems, Inc. (http://www.nimblegen.com) and consisted of 50-bp probes tiled every 20 bp on both forward and reverse strands, with no repeat masking. Three replicates for each strand were spotted on the array. Mononucleosomal DNA and genomic DNA were labeled with Cy3 and Cy5, respectively, and hybridized to the array by the manufacturer. The authors estimate that a single microarray chip constructed in this way can interrogate up to 200 kb of DNA sequence. Capillary electrophoresis-based sequencing can only cover ~10 kb at a time (Dennis et al., 2007), making it less suitable for a genome-wide application. Rather, it is a useful method for verifying nucleosome positions obtained by other, potentially less accurate, means. The authors carried out a proof-of-principle experiment by designing microarray probes that spanned two relatively short genomic regions from a human breast cancer cell line: MMTV-LTR and *LCMT2*. The study focused on the consistency of the results obtained by the two techniques, and on the good correspondence with previously published nucleosome positions in the MMTV-LTR locus (Richardfoy and Hager, 1987; Fragoso et al., 1995).

The first high-throughput sequencing study of *H. sapiens* nucleosomes was carried out in 2007 by Keji Zhao and coworkers (Barski et al., 2007). The authors employed a Solexa 1G Genome Analyzer to directly sequence ChIP DNA from mononucleosomes generated by MNase digestion of native chromatin. They collected 36-bp long sequence reads for the genome-wide distribution of nucleosomes with

20 post-translational histone modifications as well as the histone variant H2A.Z. The latter data set with 7.4 million sequence reads is analogous to the yeast and fly H2A.Z maps to the human genome (Albert et al., 2007; Mavrich et al., 2008b). Similarly to yeast, H2A.Z was found to be enriched in promoter regions both upstream and downstream of the TSS. Furthermore, H2A.Z binding correlated with gene expression, with H2A.Z-containing nucleosomes present at higher levels in promoters of active versus silent genes.

In the follow-up work published by the same group in 2008, the authors chose resting and activated human CD4+ T cells as a model system (Schones et al., 2008). As in previous studies, mononucleosome-sized DNA was isolated from MNase-digested chromatin and sequenced using the Solexa/Illumina short-read technique; 25-bp reads corresponding to the ends of ~150 bp mononucleosome cores were mapped onto the reference human genome. This procedure yielded 154,582,677 uniquely mapped reads in resting cells and 141,931,997 uniquely mapped reads in activated cells. Only those reads were retained for further analysis. The authors found a familiar pattern of phased nucleosomes around the TSS – similarly to yeast and fly, the +1 and –1 nucleosomes (–1 nucleosome is labeled –2 in Schones et al., 2008) flank an NDR and serve as termini of nucleosome arrays that become progressively less phased with the distance to the TSS.

Nucleosome phasing with respect to TSS was found to be more pronounced in expressed rather than unexpressed genes, consistent with the earlier observation in yeast showing the absence of nucleosome phasing or depletion in a cluster enriched for stress response genes (Lee et al., 2007) (Stress response genes are not expressed in the YPD medium used to prepare the cell culture in the Lee et al. study.) The authors observed that the position of the +1 nucleosome depends on gene expression: its 5′ end peaked at +40 bp with respect to the TSS in active promoters, but only at +10 bp in inactive promoters. This may be due to Pol II binding in the promoter regions of active genes: the Pol II peak was found to be located around +10 bp (Barski et al., 2007), overlapping with the nucleosome peak in inactive promoters. Finally, the authors found that gene activation by T cell receptor signaling was accompanied to some extent by nucleosome reorganization in promoters and enhancers: there is a consistent difference in the nucleosome occupancies of resting and activated cells for induced and repressed genes.

B. Structural Studies of the Nucleosome Core Particle

As of 2009, there were approximately 25 nucleosome structures with resolution ≤3.0 Å available in the Protein Databank (PDB; www.rcsb.org). Although the histones in these structures are derived from chicken, mouse, human, yeast, and frog, most of them contain the same 146-bp sequence of α-satellite DNA with nearly identical geometries. Several exceptions include a 145-bp sequence (PDB code 2nzd), sequences that are two mutations away from the α-satellite sequences (2cv5, 1kx3, 1aoi), and a 147-bp highest-resolution (1.9 Å) entry with a single base pair insertion (1kx5) (Richmond and Davey, 2003). In addition, 2fj7 (solved at 3.2 Å resolution) contains a 16-bp poly(dA-dT) tract in its 147-bp long DNA sequence (Bao et al., 2006).

There are also several structures with additions and variations, including three nucleosomes derived from 1aoi and complexed with site-specific minor groove-binding ligands, the pyrrole-imidazole polyamides (Suto et al., 2003) (1m18, 1m19, 1m1a); crystal structures with histone Sin mutants (Muthurajan et al., 2004) (the 1p3 series), and histone variants H2A.Z (Suto et al., 2000) (1f66) and macroH2A (Chakravarthy et al., 2005) (1u35); and a structure with a pyrrole-imidazole hairpin polyamide that spans the nucleosomal "supergroove" (Edayathumangalam et al., 2004) (1s32).

As can be seen from the above catalog, the vast space of nucleosomal DNA conformations and sequences has not yet been sufficiently sampled by structural studies. Thus understanding the rules linking DNA geometries with nucleosome positioning and free energies of nucleosome formation requires computational modeling guided by available structural data. In particular, the 1.9 Å crystal structure of the nucleosome core particle (Richmond and Davey, 2003) reveals the details of a single DNA conformation with unprecedented accuracy. The DNA structure is remarkably different from that observed in non-histone protein–DNA complexes: DNA trajectory has more than twice the curvature of the ideal superhelix with the same radius (41.9 Å) and pitch (25.9 Å) (Fig. 6a). Thus DNA is strongly bent, kinked, and twisted on average, with DNA segments bent into the minor groove either kinked or alternatively shifted. This is not surprising because DNA is stiff at the length scales of a single core particle and therefore bending it into a superhelix requires major conformational distortions of free B-DNA.

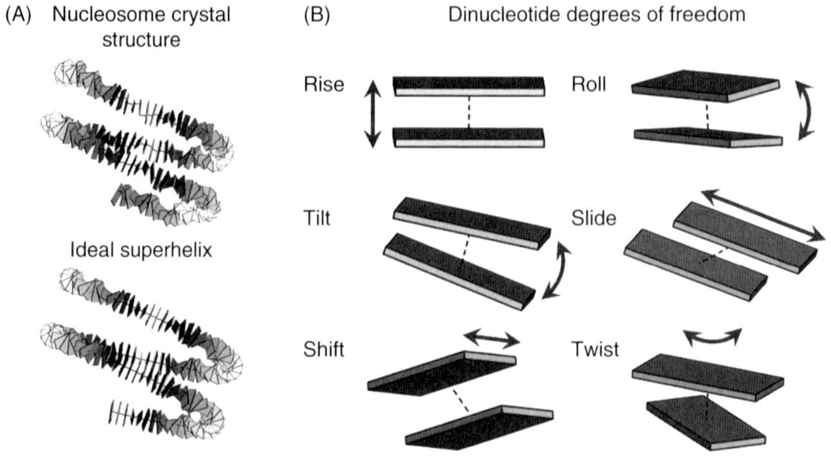

FIG. 6. (A) DNA conformation from the crystal structure of the nucleosome core particle (PDB code 1kx5) (Richmond and Davey, 2003) and the ideal superhelix. (B) Conformation of a single dinucleotide (defined as two consecutive DNA base pairs in the 5′–3′ direction) is described by six geometric degrees of freedom: three relative displacements (rise, shift, and slide) and three angles (twist, roll, and tilt). DNA base pairs are shown as rectangular blocks, and the direction of each displacement or rotation is indicated with arrows.

III. Computational Studies of Chromatin Structure

A. *Using DNA Elasticity Theory to Predict Nucleosome Formation Energies*

The availability of crystal- and NMR structures of nucleosome core particles and other protein–DNA complexes makes it possible to predict free energies of nucleosome formation and nucleosome occupancy profiles *ab initio*, without resorting to high-throughput nucleosome-positioning data sets described in Section II.A. When a nucleosome core particle is formed, a 147-bp long DNA molecule wraps around the surface of the histone octamer in ~1.65 turns of a left-handed superhelix (Richmond and Davey, 2003). Because DNA wraps so tightly around the histone core (the length of nucleosomal DNA is comparable to the DNA persistence length), the free energy of bending DNA into a superhelical shape is strongly sequence-dependent: flexible sequences bend more easily than rigid DNA molecules such as poly(dA-dT) tracts. It is reasonable to assume that DNA bending depends mostly on base-stacking energies, so that the total free energy is given by the sum of the individual dinucleotide contributions.

Typically, DNA base-stacking geometries are defined using three relative displacements (rise, shift, and slide) and three rotation angles (twist, roll, and tilt) for each dinucleotide formed by two adjacent base pairs (Fig. 6b) (Olson et al., 1998). Together the six degrees of freedom completely specify the spatial position of base pair $i+1$ in the local coordinate frame attached to base pair i. Cartesian coordinates of an arbitrary DNA molecule can be used to construct a full set of relative dinucleotide geometries. Conversely, specifying a complete set of dinucleotide degrees of freedom is sufficient for reconstructing an arbitrary DNA conformation in global Cartesian coordinates (Lu et al., 1997a, 1997b).

As shown by Olson et al. (1998), these degrees of freedom can be used to derive an empirical model of DNA elastic energies. The model takes distributions of dinucleotide geometries observed in the ensemble of non-homologous protein–DNA structures as input (there are currently >100 such structures in the Protein Data Bank). These data are used to compute the mean and the covariance matrix for each degree of freedom (rise, shift, ...) and for each dinucleotide type (AA, AC, ...). Retaining the full covariance matrix allows the model to include correlations between different degrees of freedom. DNA elastic energy is described by an effective quadratic potential:

$$E_{\text{el}} = \frac{1}{2}\sum_{s=1}^{N}[\alpha^s - \langle\alpha^{n(s)}\rangle]^T F^{n(s)}[\alpha^s - \langle\alpha^{n(s)}\rangle], \tag{1}$$

where α^s is the six-component vector of dinucleotide degrees of freedom; the sum runs over $N = 146$ consecutive dinucleotides; and $\langle\alpha^{n(s)}\rangle$ is the vector of average values for each degree of freedom for the dinucleotide of type n at position s. $F^{n(s)}$ is a matrix of stiffness coefficients computed by inverting the covariance matrix C^n for the dinucleotide of type n at position s: $F^n = (C^n)^{-1}$, where

$$C_{ij}^n = \langle(\alpha_i^n - \langle\alpha_i^n\rangle)(\alpha_j^n - \langle\alpha_j^n\rangle)\rangle. \tag{2}$$

Note that the elastic energy model utilizes only the first and second moments of the empirical distributions of dinucleotide geometries – the available structural data is insufficient for retaining higher order moments or for modeling more than two consecutive base pairs.

The elastic energy model described above was first adapted to predicting nucleosome positions in a 2007 paper by Tolstorukov et al. (2007). It is known from the 1.9 Å resolution crystal structure of the nucleosome core particle (Richmond and Davey, 2003) that dinucleotide positions at which nucleosomal DNA is kinked have large positive values of slide and large negative values of roll (solid lines with squares in Fig. 7). The kinks mediate bending DNA into the nucleosomal shape and define "hot spots" at which high-affinity NPSs have flexible dinucleotides.

The authors argue that lateral slide deformations observed at sites of local anisotropic bending define the superhelical trajectory of nucleosomal DNA. They show that slide accounts for over 90% of the overall pitch of nucleosomal DNA, with stepwise accumulation of net pitch in ~10–11 bp increments. The positive values of slide accompany DNA bending into the minor groove where roll is negative, and the negative values of slide appear with DNA bending into the major groove where roll is positive (Fig. 7). Because the direction of slide is roughly parallel to the superhelical axis at the kink sites, the values of slide accumulate along the path of nucleosomal DNA. As a result, steps with alternating positive and negative slide (separated by ~5–6 bp) contribute cooperatively to the overall superhelical pitch.

While Tolstorukov et al. mostly focused on the contributions of roll and slide to the nucleosomal geometry, they also used DNA elastic energies to predict optimal positions of mononucleosomes reconstituted *in vitro* on four short DNA sequences and mapped by hydroxyl radical footprinting. Elastic energy was computed for each allowed position of the nucleosome core particle along each DNA segment, and energy minima were identified with predicted optimal positions. DNA geometry was taken from the high-resolution crystal structure (Richmond and Davey, 2003), under the *ad hoc* assumption that sequence-dependent variations in DNA geometries and kink positions were not important for predicting nucleosomal energies. Regardless of this limitation, the model showed reasonable predictive power, although it is difficult to judge how its performance would scale up in genome-wide calculations where multiple nucleosomes form simultaneously under steric constraints.

Another approach utilizing the empirical DNA elastic potential was developed by Morozov et al. (2009). The main methodological difference from Tolstorukov et al. (2007) is that the DNA geometry was allowed to be relaxed from its initial conformation (given *e.g.* by the ideal superhelix

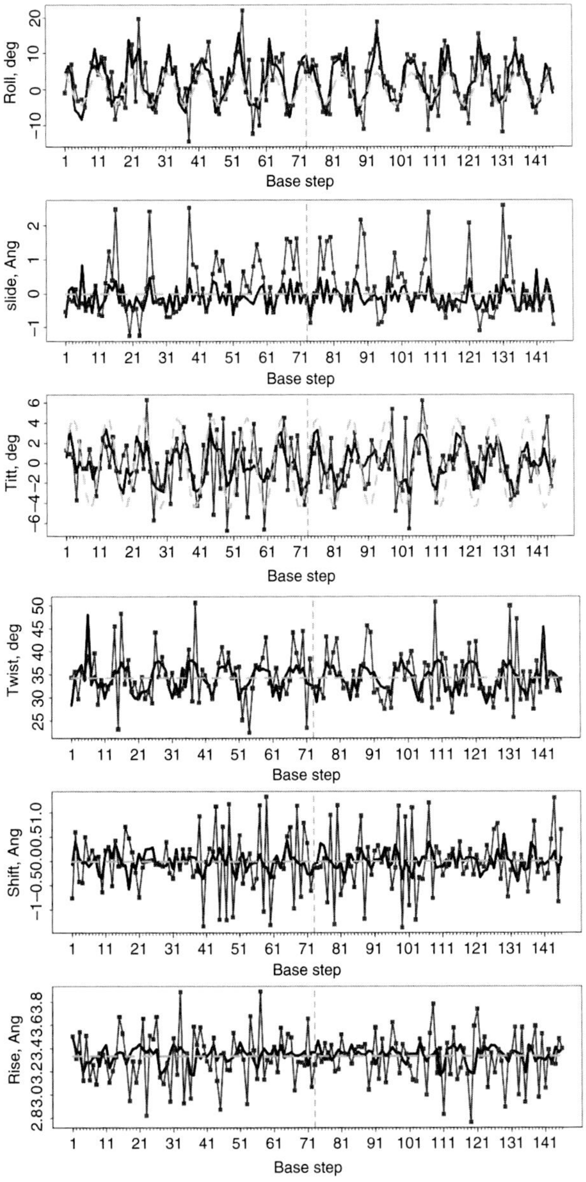

FIG. 7. (Continued)

or by the nucleosome crystal structure). Thus the total energy of a nucleosomal DNA was given by a weighted sum of two quadratic potentials:

$$E = E_{\mathrm{el}} + wE_{\mathrm{sh}}, \tag{3}$$

where E_{el} is the sequence-dependent DNA elastic potential and E_{sh} is the non-specific histone–DNA interaction energy designed to penalize deviations of nucleosomal DNA from the ideal superhelix:

$$E_{\mathrm{sh}} = \sum_{s=1}^{N} (\vec{r}_s - \vec{r}_s^{\,0})^2, \tag{4}$$

where $\vec{r}_s$ and $\vec{r}_s^{\,0}$ are the nucleosomal DNA and the ideal superhelix radius vectors to the origin of base pair s in the global frame (the base pair origin is defined by its atomic coordinates Lu et al., 1997a, 1997b). Though this term is an oversimplification of complex atomic interactions between histones and DNA, its quadratic form reduces minimization of the total energy E to solving a system of linear equations: "The weight w is set to 0.1 to optimize DNA geometry predictions." The final conformation of the DNA molecule is then the one that minimizes its total energy.

Using this approach the authors carried out the first *ab initio* prediction of DNA base step geometries in nucleosomal DNA. With only the 147-bp α-satellite sequence from 1kx5 as input, they predicted 6 dinucleotide degrees of freedom observed in the crystal structure with the average correlation coefficient of 0.46 (solid lines in Fig. 7). In comparison, the average correlation coefficient between the ideal superhelical geometry and 1kx5 was 0.07. The model underpredicted the absolute values of

FIG. 7. Dinucleotide geometries from the crystal structure of the nucleosome core particle (solid red lines with squares, PDB code 1kx5) (Richmond and Davey, 2003): from the minimum-energy structure obtained by relaxing the 147-bp DNA segment with the nucleotide sequence from 1kx5 around the ideal superhelix as described in Section III.A (solid black lines) (Morozov et al., 2009); and from the ideal superhelix (dashed cyan lines). Six dinucleotide degrees of freedom are shown: roll, slide, tilt, twist, shift, and rise. The twofold nucleosome symmetry axis is shown as a dashed vertical line. Note that roll is negative when DNA bends into the minor groove (which faces the histone octamer) and positive when DNA bends into the major groove (with the minor groove facing away from the histone octamer). (See color plate 3) slide in kink positions (although the overall correlation was above average at 0.54) and did not reproduce rapid alternate shifts between base pairs 31 and 111 (Fig. 7). This behavior could be due to inaccuracies in the coefficients of the empirical elastic potential (derived from non-histone protein–DNA complexes) or due to the inherent limitations of the quadratic model (Eq. (3)).

The authors also predicted free energies of nucleosome formation for ~45 *in vitro* sequences (with a correlation coefficient of ~0.8), as well as nucleosome positions for six short sequences with nucleosomes mapped by hydroxyl radical footprinting (four from Tolstorukov et al. (2007) and two more determined *de novo*). The authors found that DNA geometry relaxation helped with predicting nucleosome free energies but had surprisingly little effect on the accuracy of predicted nucleosome positions.

Finally, the authors provided an exact numerical solution to the many-body problem of placing multiple nucleosomes onto longer stretches of DNA ((Durbin et al., 1998; Segal et al., 2006); see Section III.C for details). Free energies of nucleosome formation at every DNA base pair were translated by the algorithm into nucleosome probabilities and occupancies under the assumption of steric exclusion among nucleosome core particles of fixed size.

Miele et al. (2008) have also used a sequence-dependent DNA flexibility model to predict nucleosome occupancies in *S. cerevisiae* and *D. melanogaster* genomes. Their model employed the ideal superhelical geometry (with pitch and radius from 1kx5) and considered only the angular (roll, twist, and tilt) isotropic contributions to the superhelical curvature. In other words, DNA was modeled as an inextensible and unshearable elastic rod, and correlations between different degrees of freedom were neglected. The equilibrium values of the angular parameters and the stiffness coefficients were adopted from Anselmi et al. (2000). These parameters are not based on protein–DNA structural data – rather, they were derived by energy calculations in the framework of the nearest-neighbor approximation (De Santis et al., 1986) and later refined to improve the correlation between the calculated and the experimental gel electrophoresis mobility of a large pool of synthetic and natural DNA molecules. The authors also took into account the entropic cost of the transition from free to superhelical DNA, and computed the free energy difference ΔF at every base pair along the sequence.

In yeast, the free energy landscape was compared with the log2 ratio of hybridization values from Yuan et al. (2005) and Lee et al. (2007). The authors observed significant correlation between their predictions and experimental data. In particular, the model was able to discriminate between sets of DNA fragments with the highest and the lowest nucleosome occupancies (with the area under the receiver operating characteristic (ROC) curve being 0.72 for the data set from Yuan et al.). Furthermore, the model predicted 5′ NDRs. In the fly, the model exhibited nucleosome occupancy depletion (*i.e.* higher values of ΔF) in the vicinity of target sites of the Trithorax and Polycomb group proteins zeste and Ez/Psc. This depletion was also observed in the experimental profile of chromatin sensitivity to MNase obtained in a study of histone H3.3 replacement patterns in *D. melanogaster* (Mito et al., 2007).

DNA elastic parameters used by Miele et al. were also employed for nucleosome positioning predictions by the original authors of the force field, De Santis and colleagues (Anselmi et al., 2000, 2002; Scipioni et al., 2009). In addition to the DNA model based on roll, tilt, and twist, the conformational entropy contribution to the relative thermodynamic stability of the nucleosome core particle is explicitly considered by the authors and an empirical term that depends on the curvature of free DNA is introduced. To justify the latter term, the authors observe that DNA elastic energies alone give a satisfactory agreement for relatively straight DNA but exhibit large deviations for intrinsically curved DNA. The magnitude of these deviations shows a strong correlation with $\langle A_f^0 \rangle$, the average integral curvature of free DNA. On the basis of this observation the authors added an empirical contribution of $4.5\langle A_f^0 \rangle^{3/2}$ to the model.

The De Santis et al. approach leads to accurate predictions of free energies of nucleosome formation ($r = 0.92$ on a set of ~100 mononucleosomal sequences collected from the literature) (Scipioni et al., 2009). The authors also compute free-energy profiles for several genomic loci and compare them with experimental data from Yuan et al. (2005) and Kaplan et al. (2009). Unfortunately, the authors do not provide overall measures of performance such as ROC curves, making it difficult to compare their approach with other methods.

Another model that considers DNA to be an elastic rod is due to Vaillant et al. (2007). Although DNA is formally described by all three local angles (roll, tilt, and twist), only the roll degree of freedom is sequence-dependent, while the equilibrium value of tilt is set to 0 and the equilibrium value of twist is set to $2\pi/10.5$ for all dinucleotides. Because the geometry of nucleosomal DNA is assumed to be described by an ideal superhelix, only the roll degree of freedom contributes to the overall elastic energy (cf. Eq. (1)). Equilibrium values of roll are based on a trinucleotide coding table from Goodsell and Dickerson (Goodsell and Dickerson, 1994), who in turn adopt it from Satchwell et al. (1986). The coding table uses a relatively small set of NPSs to compute the fractional preference of each base pair triplet to be outside or inside of the DNA wound around the histone octamer. "Outside" and "inside" refer to the position of the major groove with respect to the histone surface: for example, the GGC triplet has a 45% preference for locations on a bent double helix in which its major groove faces inward and is compressed by the curvature (corresponding to the region of the positive roll), whereas the AAA triplet has a 36% preference for the opposite orientation, with the major groove facing outward (corresponding to the region of the negative roll).

Despite being phrased in the language of DNA elastic energies, the Vaillant et al. model has more in common with statistical nucleosome positioning scores described in Section III.B (Ioshikhes et al., 2006; Segal et al., 2006; Field et al., 2008; Kaplan et al., 2009) than with physical models. The roll degree of freedom is simply used to quantify the relative orientational preferences of trinucleotides with respect to the helical twist and could have been replaced by position-dependent scores without any explicit reference to DNA geometry.

B. *Bioinformatics Models of Nucleosome Sequence Preferences*

An alternative approach to predicting nucleosome positions is based on training bioinformatics models to discriminate various sequence features that differentiate nucleosomal and linker DNA. This approach requires extensive collections of nucleosomal sequences available from recent high-throughput nucleosome positioning experiments described in Section II.A (we note in passing that two "first-generation" nucleosome positioning models, Ioshikes et al. (2006) and Segal et al. (2006), used much smaller, ~200, collections of NPSs. These early approaches have been largely superseded by the more recent models trained on high-throughput data.)

1. Segal et al. Model Based on Dinucleotide Distributions and 5-mer Counts

One bioinformatics model that positions nucleosomes according to their sequence preferences was developed in Field et al. (2008) and Kaplan et al. (2009). This model assigns a nucleosome formation score to each 147-bp long DNA sequence. The score accounts for two major nucleosome positioning signals: the higher affinity of histone octamers for periodic distributions of certain dinucleotides due to anisotropic sequence-dependent DNA bending (rotational positioning), and longer motifs that function as nucleosome favoring or disfavoring signals, such as poly(dA-dT) tracts found mainly in the nucleosome-depleted regions (translational positioning). Similarly to DNA elastic energies, the bioinformatics score can be used to evaluate genome-wide nucleosome occupancies with steric exclusion, as described in Section III.C below. The latest iteration of this model is trained on the *in vitro* high-throughput sequencing data set discussed in Section II.A.1.2, and is capable of predicting genome-wide nucleosome occupancies with a correlation coefficient of 0.89 for *in vitro* data and 0.75 for *in vivo* data (Kaplan et al., 2009).

As mentioned above, the Segal et al. model consists of two components: P_N and P_L. The first component, P_N, captures position-dependent periodicity of dinucleotide distributions (Fig. 5). In analogy with position-specific scoring matrices (Stormo and Fields, 1998), the nucleosome-bound DNA sequences and their reverse complements are aligned around the dyad axis. The alignment is used to estimate the conditional dinucleotide distribution $P_{N,i}(S_i|S_{i-1})$ for each $i=2,\ldots,147$ – that is, the probability of observing the nucleotide S_i at position i given the nucleotide S_{i-1} at position $i-1$. The first component of the model is then defined as:

$$P_N(S) = P_{N,1}(S_1)\prod_{i=2}^{147} P_{N,i}(S_i|S_{i-1}). \quad (5)$$

Thus to any DNA sequence S of length 147 bp, one can assign a probability using Eq. (5). As this component captures nucleosomal dinucleotide sequence preferences, the probability is higher for sequences with ~10–11 bp phased patterns of A/T- and G/C-rich dinucleotides (Fig. 5).

The second component of the score, P_L, accounts for the difference in global, position-independent distributions of 5 bp long words in nucleosomal and linker sequences (5-mers are chosen here for computational reasons). To find out which of the 1,024 sequences of length 5 is favored or disfavored by nucleosomes, probability P_l is evaluated for each word of length 5 as the ratio between the frequency of that word in linkers and its frequency in the nucleosomal sequences. This ratio is normalized by the sum of all ratios across all words of length 5. So, by definition, sequences with higher probability will be less favorable for nucleosomes than sequences with lower probability. The position-independent component $P_L(S)$ for a given sequence S of length 147 is then defined as a product over all instances of 5-mers in S:

$$P_L(S) = \prod_{i=5}^{147} P_l(S_i|S_{i-4}, \ldots, S_{i-1}). \tag{6}$$

By construction P_L is higher for nucleosome-depleted sequences, in contrast to P_N, which is higher for nucleosome-bound sequences. Both of these components are used to predict nucleosome occupancies genome-wide (Field et al., 2008; Kaplan et al., 2009; Segal and Widom, 2009): P_N captures rotational positioning, while P_L describes translational positioning. The log-ratio between the components

$$E(S) = -\log \frac{P_N(S)}{P_L(S)}, \tag{7}$$

assigns a nucleosome formation "energy" to each sequence S of length 147 bp. Eq. (7) scores are then used to formulate a thermodynamic model for predicting nucleosome occupancies and positions (see Section III.C).

2. *Ioshikes et al. Comparative Genomics Model*

Another computational model that employed position-dependent periodicity of nucleosomal dinucleotide distributions was developed by

Ioshikhes et al. (2006). The authors used a periodic 139-bp dinucleotide AA/TT signal as an empirical definition of an NPS. The periodic pattern used in this work was compiled by Ioshikhes et al. in a previous study (Ioshikhes et al., 1996), where a collection of 204 nucleosomal DNA sequences was used to compute the AA/TT positional frequency distributions. The substantial noise in the data had been reduced by employing five different multiple sequence alignment techniques with subsequent averaging of the AA/TT positional frequency profiles.

In Ioshikhes et al. (2006) the entire *S. cerevisiae* genome was scanned for correlations with the NPS. To increase the method's resolution the authors utilized a comparative genomics approach; specifically, they calculated NPS correlation profiles for each gene between −1,000 bp and +800 bp, relative to the +1 ATG start codon, in six sequenced *Saccharomyces* species. The authors argued that well-positioned nucleosomes should be conserved at orthologous locations in all related species. Thus averaging across orthologous locations in different *Saccharomyces* species would suppress noise and increase the sensitivity of the method compared to a single-genome approach. The Ioshikhes et al. study revealed several nucleosome positioning features conserved across related yeast species. Specifically, most yeast promoters have a nucleosome-depleted region, revealed by the most negative correlation between the underlying DNA sequence and the periodic NPS. Well-positioned +1 nucleosomes just downstream of the NDR (Fig. 3) exhibit maximum positive correlation between the underlying DNA sequence and the NPS. The authors also pointed out a difference in correlation patterns for TATA-less and TATA-containing genes: TATA-less genes tend to have a more consistent NPS–NDR–NPS pattern.

3. *Support Vector Machine for Identifying the Nucleosome Formation Potential*

A discriminative approach to predicting nucleosome positions was developed by Peckham et al. (2007). Specifically, Pekham et al. implemented a support vector machine (SVM) (Vapnik, 1998) to distinguish between nucleosome-forming and nucleosome-inhibiting DNA sequences. Because the SVM algorithm performs discriminative data classification in vector space, Pekham et al. converted each 50-bp long DNA sequence into a vector of *k*-mer frequencies, where *k* runs from 1 to 6. By means of this procedure every DNA sequence can be uniquely mapped into a 2,772-dimensional vector space. The SVM algorithm finds a hyperplane in this vector space that separates two groups of training data in such a way that the distance from the hyperplane to the nearest data point is maximized.

The training data used by Pekham et al. was taken from a low-resolution microarray study by Yuan et al. (2005) (see Section II.A.1.1), where nucleosome occupancies were measured for yeast chromosome III. The training set consisted of 1,000 nucleosome-forming (50 bp long) DNA sequences with the highest hybridization scores and 1,000 nucleosome-inhibiting (50 bp long) DNA sequences with the lowest hybridization scores. An SVM trained on this data set can classify DNA segments as nucleosome-forming or nucleosome-inhibiting after mapping them into the vector space.

The authors employed the trained SVM to estimate the nucleosome formation potential of a collection of ~200 nucleosomal DNA sequences studied in Segal et al. (2006). The nucleosome formation potential correlated with the dinucleotide AA/TT/AT periodic signal found by Segal et al. (2006). The authors argued that G/C and A/T content of a sequence is the strongest predictive factor in determining the nucleosome formation potential: G/C-rich sequences favor nucleosome formation, whereas A/T-rich sequences disfavor it. They also pointed out that A-tracts tend to prevent nucleosome formation. Based on how well the SVM discriminated between nucleosome-bound and nucleosome-free genomic sequences, the authors concluded that only ~50% of all nucleosomes are positioned by intrinsic DNA sequence signals.

4. Wavelet-Based Approach to Discriminating Nucleosome-Bound from Nucleosome-Free Sequences

Yuan and Liu (2008) proposed an alternative computational model for predicting nucleosome positions, based on a wavelet transform. Similarly to Peckham et al. (2007), the authors developed a classification algorithm that distinguishes nucleosome-bound sequences from linker DNA. Each DNA sequence S of length 131 bp is decomposed into 16 numerical vectors of length 130 bp, where each vector describes one of 16 dinucleotides. The ith component of this vector is 1 if the corresponding dinucleotide is found at location i in sequence S, and 0 otherwise. The moving average with a 3 bp window is applied to the vectors resulting in sequences of length 128 bp (the length $128 = 2^7$ is motivated by the discrete wavelet transform). Thus each of the 16 vectors holds information about the corresponding dinucleotide frequency in *S*.

The method used by Yuan and Liu (2008) is based on a wavelet transform which allows to detect periodic patterns in the signal over multiple scales (Mallat, 1999). In the wavelet analysis the signal is decomposed into orthogonal components corresponding to different frequency bands. The advantage of this approach over Fourier spectral analysis is that the wavelet transformation captures not only the frequency component but also the position of this component in the signal. The coefficients of the wavelet transform characterize periodic patterns embedded in the dinucleotide frequency signal.

Thus, if a certain frequency is associated with a putative NPS, it can be detected by comparing the contributions (defined by the coefficients of the wavelet transform) from nucleosomal versus linker DNA sequences. The probability for a DNA sequence *S* to be a nucleosomal sequence is then defined through a logistic regression model with the co-variates given by the coefficients of the discrete wavelet transform. The model was trained on previously identified nucleosome and linker sequences (Yuan et al., 2005; Segal et al., 2006). In contrast to the Pekham et al. (2007) model, this approach only accounts for dinucleotide frequencies. The authors conclude that sequence information is highly predictive of local nucleosome enrichment or depletion, whereas predictions of the exact nucleosome positions are only moderately accurate, suggesting the importance of other regulatory factors in fine-tuning nucleosome positions.

C. *Statistical Physics of One-Dimensional Liquids and the Nucleosome Positioning Problem*

The recent high-resolution maps of nucleosome locations in eukaryotic genomes (Lee et al., 2007; Mavrich et al., 2008a, 2008b; Schones et al., 2008; Shivaswamy et al., 2008; Valouev et al., 2008) reveal that nucleosomes are arranged in quasi-periodic arrays, covering ~70–80% of genomic DNA. Furthermore, regions upstream of open reading frames (ORFs) are typically depleted of nucleosomes. Nucleosome-depleted promoter regions are flanked on both sides by the so-called ±1 nucleosomes (Jiang and Pugh, 2009). It is believed (Mavrich et al., 2008a) that these nucleosomes act as boundaries, inducing periodic oscillations in the neighboring nucleosome occupancy (Fig. 3). These oscillations are caused by steric exclusion and can be explained with a simple statistical model of a one-dimensional (1D) liquid. In this model the nucleosomes are treated as a uniform 1D liquid of 147-bp hard rods. The simplest approach of this kind assumes that nucleosomes have no intrinsic sequence specificity and are positioned solely by steric exclusion (Kornberg and Stryer, 1988). To induce periodic oscillations in the nucleosome occupancy, a boundary constraint is introduced at some point along the DNA. The boundary may be due to the sequence-specific ± 1 nucleosomes or to DNA-bound non-histone proteins.

Nucleosome reconstitution *in vitro* involves stepwise dialysis from concentrated salt solutions (when DNA molecules have low affinity for the histone octamer due to electrostatic screening) down to physiological salt concentrations (Shrader and Crothers, 1989, 1990; Lowary and Widom, 1997). This is expected to lead to an equilibrium distribution of nucleosomes, which can be described by equilibrium statistical mechanics. Nucleosome positions *in vivo* are influenced by the presence of chromatin remodelers and TFs, and many other factors that in principle may perturb the system out of equilibrium. However, Kaplan et al. found that correlation between *in vitro* and *in vivo* occupancy profiles is quite high, more than 70% on average (Kaplan et al., 2009). Thus both *in vivo* and *in vitro* chromatin can be described using equilibrium methods.

In the Kornberg and Stryer model nucleosomal arrays can be regarded as an ensemble of two kinds of objects: nucleosomes of length a bp and DNA linkers with a mean length of L bp. The probability of selecting a nucleosome from the ensemble of a single nucleosome and L linkers of unit length is $p = (1 + L)^{-1}$, and the probability of selecting a linker is $1 - p$. A stretch of DNA of length x can contain any number of nucleosomes between 0 and x/a. Let $M \equiv \mathrm{int}(x/a)$ be the maximum number of nucleosomes that can still fit into the DNA domain of length x. The partition function for this domain is then defined as a binomial sum over all possible configurations of nucleosomes and linkers:

$$\Xi(x) = \sum_{n=0}^{M} \binom{n + x - an}{n} p^n (1 - p)^{x-an}. \tag{8}$$

Kornberg and Stryer used this model to demonstrate how nucleosomes without intrinsic sequence preferences can nonetheless be organized into periodic arrays (Kornberg and Stryer, 1988). Equation (8) is used to calculate the probability of any configuration of nucleosomes and linkers on a DNA molecule of arbitrary length N. The probability $P(i)$ that site i is occupied by a linker (*i.e.* is nucleosome-free) is derived by noting that the linker site i divides the DNA molecule into two regions with lengths i–1 and N–i and the linker site itself:

$$P(i) = \Xi(i-1)\frac{1-p}{\Xi(N)}\Xi(N-i). \tag{9}$$

The normalization factor $\Xi^{-1}(N)$ corresponds to all possible configurations of nucleosomes and linkers over the entire DNA molecule.

The linker probability profiles (Eq. (9)) evaluated for various values of L are shown in Fig. 8. We observe a regular spacing of nucleosomes near the boundaries with an oscillation period of $a + L$. The oscillations decay with

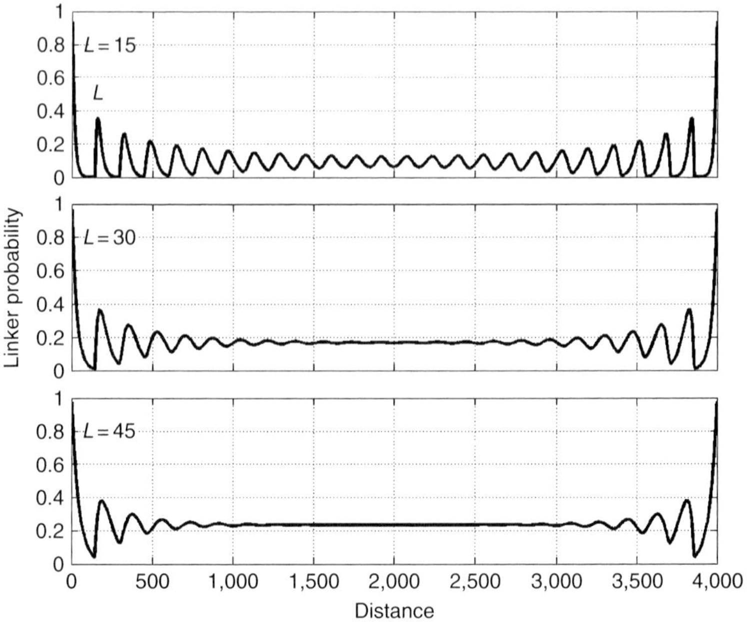

FIG. 8. Statistical positioning of nucleosomes by the boundaries. Linker probabilities computed using Eq. (9) with $N = 4000$, $a = 147$, and three different average linker lengths L are shown. The period of the oscillations depends on the linker length as $a + L$.

the distance from the boundary. Thus the boundary constraint alone leads to an array of regularly spaced nucleosomes at non-random locations. However, the significance of such boundaries in genomes and their molecular nature are unclear. The nucleosome positioning boundaries may be defined by sequence-specific DNA-bound proteins (Fedor et al., 1988; Roth et al., 1990; Pazin et al., 1997) or by rigid DNA sequences such as Poly(dA-dT) tracts that tend to exclude nucleosomes (Iyer and Struhl, 1995; Suter et al., 2000; Anderson and Widom, 2001; Bao et al., 2006; Lee et al., 2007; Mavrich et al., 2008a; Segal and Widom, 2009).

In the Kornberg and Stryer statistical model nucleosome positions are dictated solely by steric constraints. Such a model does not account for histone–DNA interactions, which should introduce a certain amount of sequence specificity into nucleosome positioning on DNA. However, it is straightforward to extend the model to the sequence-specific case and thus predict nucleosome occupancies for histone–DNA interactions of arbitrary magnitude.

For DNA sequence S and a set of k nucleosomes at positions $n_1, n_2, \ldots, n_k$, one defines a statistical weight function as a product of the corresponding Boltzmann factors:

$$W[S,n] = \prod_{i=1}^{k} \tau \exp\{-\beta E(S_{n_i,n_i+146})\}, \tag{10}$$

where n denotes the specific configuration of nucleosomes, β is the inverse temperature, S_{n_i,n_i+146} is the part of sequence S covered by the nucleosome n_i, and τ is the apparent nucleosome concentration (Field et al., 2008). The nucleosome positions $n_1, n_2, \ldots n_k$ are chosen in such a way that no two nucleosomes overlap. To find the probability of a given configuration n of nucleosomes on S the statistical weight is divided by a partition function, which is the sum over all allowed nucleosome configurations: $P(W[S,n]) = W[S,n]/\Sigma_{n'} W[S,n']$.

The number of allowed configurations for long DNA sequences is exponentially large. Nevertheless, the probability of starting a nucleosome at every genomic base pair can be efficiently computed using the dynamic programming method (Durbin et al., 1998). The idea behind the method is similar to that used in deriving Eq. (9): the nucleosome that starts at bp i divides DNA sequence S of length N into three subsequences $S_{1,i-1}$, $S_{i,i+146}$, and $S_{i+147,N}$. In the forward step one finds the partial partition function F_{i-1}, which is the sum of the nucleosome weight functions over all possible configurations on $S_{1,i-1}$. The partition function R_{i+147} corresponding to the sequence $S_{i+147,N}$ is found in the backward step. The forward and backward steps are further explained in Fig. 9. The probability that a nucleosome starts at bp i of DNA sequence S is then given by

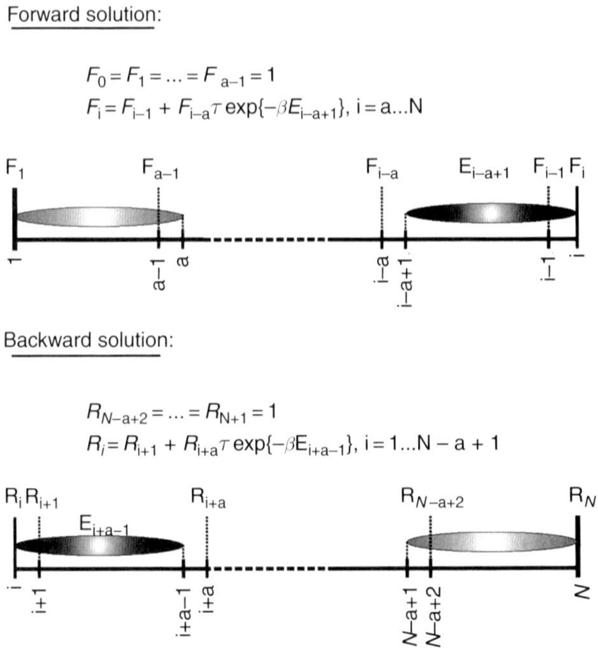

FIG. 9. Schematic representation of the forward and backward steps (Eq. (11)) in the determination of possible nucleosome configurations. The positions of nucleosomes of size a close to the boundary (light ovals) are restricted: no nucleosome can end at positions 1... $a-1$ or start at positions $N-a+2...N$. This leads to a set of boundary conditions for forward and backward partial partition functions. The forward partition function F_i corresponds to all possible configurations of nucleosomes in the interval from 1 to i. If $i \geq a$ there exist two possibilities: (1) position i is empty; this corresponds to the first term F_{i-1}, which indicates that the nucleosome configuration is unchanged from $i-1$ to i; (2) a nucleosome that started at position $i-a+1$ ends at position i (dark ovals); this case is accounted for by the second term in the recursive equation, which is the product of the nucleosome Boltzmann factor and the partition function F_{i-a} for the interval $[1...i-a]$. The backward partition function R_i is evaluated in the opposite direction and describes nucleosome configurations in the interval from i to N.

$$P_i = \frac{F_{i-1}\tau\exp\{-E(S_{i,i+146})\}R_{i+147}}{R_1}, \quad (11)$$

where $R_1 = F_N$ in the denominator is the complete partition function. The corresponding nucleosome occupancy O_i is found as the probability that a given bp i is covered by any nucleosome, and is defined as the sum of probabilities from P_{i-146} to P_i.

Although steric exclusion is taken into account in the probability profile P_i, this profile is derived for a specific value of the chemical potential μ, which enters the model through the nucleosome concentration $\tau = \exp(\beta\mu)$ in Eq. (11). Variation of this parameter leads to different probability and occupancy profiles due to the competition between the nucleosome formation energy and the excluded volume interaction. The resulting change in occupancy can be quite substantial, so it is reasonable to ask about the behavior of the model with respect to μ. In principle, this question can be addressed by re-running the dynamic programming algorithm for various values of τ. However, Schwab et al. (2008) used a more physical approach that employs statistical mechanics of a 1D liquid of hard rods in an arbitrary external field (Percus, 1976).

The system is assumed to be a grand canonical ensemble. The first component of the statistical model is the potential energy V_i of a nucleosome positioned at bp i and related to the probability P_i from Eq. (11) through $V_i = -k_B T_0 \log P_i$. The second component of the model is the interparticle potential responsible for steric hindrance. This model is solvable in a sense that the nucleosome distribution can be computed exactly for various values of the chemical potential μ and the temperature T_0, which specifies the characteristic energy scale (Schwab et al., 2008).

Both the energy profile V_i and the excluded volume interaction define the final disposition of nucleosomes on DNA. Due to the competition between these two terms a small change in the chemical potential or the strength of nucleosome binding can lead to repositioning of some of the nucleosomes (Schwab et al., 2008), as shown schematically in Fig. 10. Whereas positions of more stable nucleosomes remain unchanged, positions of less stable nucleosomes shift with μ. The mean number of nucleosomes $\langle N \rangle$ as a function of the chemical potential is shown in Fig. 10 for two values of temperature T/T_0. The observed behavior, which resembles

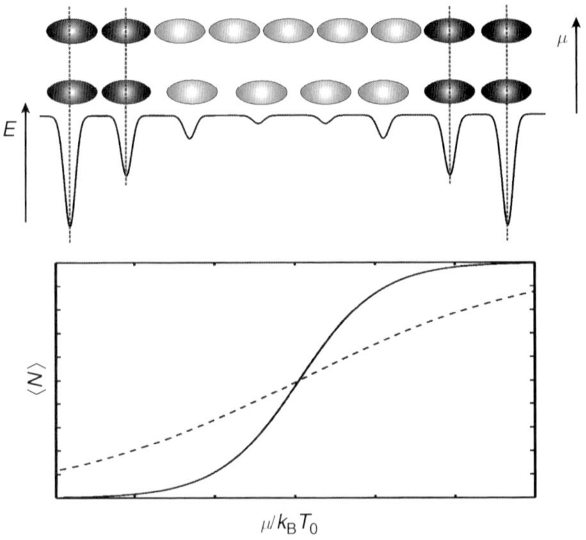

FIG. 10. Top: A nucleosome configuration is shown for two different values of the chemical potential. Stable nucleosomes (dark ovals) remain at their positions while unstable nucleosomes (light ovals) are shifted and another nucleosome appears as μ is increased. Bottom: The average number of nucleosomes on an arbitrary DNA segment is shown schematically as a function of μ for two different temperatures. Dashed line corresponds to the higher temperature.

a first-order phase transition, is a consequence of the fact that at the transition point the free energy of a given nucleosome arrangement is degenerate with respect to N. It has been demonstrated by Schwab et al. (2008) that changes in nucleosome occupancy are localized to certain regions and that the genomic locations of these regions correlate with known positions of TF-binding sites.

In the statistical models discussed here it is assumed that, apart from the steric overlap, nucleosome positions are independent and are dictated solely by intrinsic sequence preferences. In general this approximation may not hold. An internucleosomal potential can originate, for instance, from the higher order structure of the chromatin fiber. Fiber formation causes linker lengths to be quantized (Kato et al., 2003; Cohanim et al., 2006; Wang et al., 2008) because relative spatial positions of adjacent nucleosomes depend on the length of their linker DNA (Widom, 1992; Schalch et al., 2005). Another possible source of internucleosomal potential is electrostatic interactions between spatially close nucleosomes (Luger et al., 1997; Dorigo et al., 2004; Chodaparambil et al., 2007). Quantized linker lengths can be described with an effective nearest-neighbor internucleosomal potential. The thermodynamic model of Eq. (11) can be modified to account for nearest-neighbor interactions by introducing an additional term that assigns different statistical weights to different linker lengths (Lubliner and Segal, 2009).

D. Hidden Markov Models for Predicting Nucleosome Occupancies

Log-intensity profiles obtained from microarray-based nucleosome positioning experiments are often analyzed using HMMs (Rabiner, 1989). Although HMMs were originally developed for analyzing sequential stochastic signals such as noisy time-series data, they have recently found many uses in bioinformatics. In the context of chromatin structure prediction the use of HMMs was pioneered by Yuan et al. (2005) and later adopted with modifications by Lee et al. (2007) and Zawadzki et al. (2009). In each of these applications, hybridization values from the tiled array (*i.e.* log2 ratios of the nucleosomal DNA to the total genomic DNA for each microarray probe) were used as input to the HMM, which then predicted the probability of a nucleosome to start at every genomic base pair as well as nucleosome occupancy. Thus HMMs can be used to calculate nucleosome occupancies directly from the log-intensity data, in contrast to the approaches that first predict a sequence-specific nucleosome free-energy profile and then use dynamic programming to infer nucleosome positions.

HMMs are defined by the transition matrix between hidden states probabilistically assigned to each consecutive probe on the tiling microarray. For Affymetrix arrays with probes tiled at every 4 bp, 37 nucleosomal nodes and 1 linker node need to be defined, resulting in a 148-bp long nucleosome core particle (Zawadzki et al., 2009). As shown in Fig. 11A, from the linker state a transition can be made to another linker state with a probability P_{LL}, or to the first nucleosome state with a probability $P_{LN} = 1$ P_{LL}. However, once a new nucleosome is started all subsequent nodes are placed with probability 1 until the next linker state is reached.

The model used by Zawadzki et al. employs a mixture of two Gaussians to represent both the nucleosome state and the linker state (earlier models used a single Gaussian). Thus the final set of fitting parameters includes means and widths of four 1D Gaussians, two independent mixture coefficients, and 38 initial probabilities for the nucleosome and linker states. To reduce the number of fitting parameters, 37 nucleosome states were forced to share the same mixture of

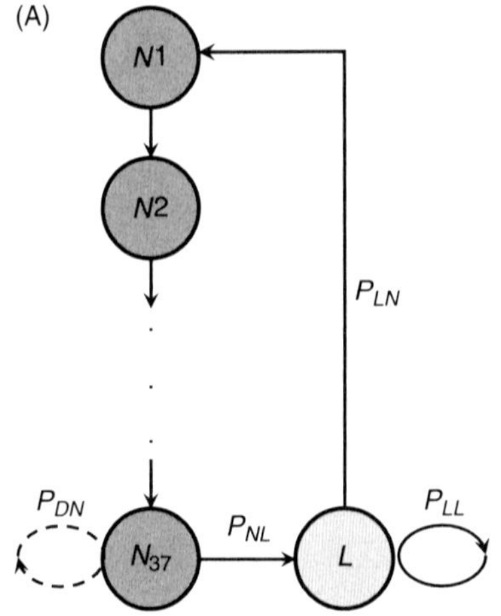

FIG. 11. (A) The transition matrix of the Hidden Markov Model (HMM) used for predicting nucleosome probabilities and occupancies from microarray log-intensity profiles. Nodes N_1 through N_{37} represent consecutive nucleosomal states (green circles), while L is the linker state (yellow circle). Once a new nucleosome is started 36 subsequent nodes are placed with a probability 1. From N_{37} a transition is made to the linker state with a probability P_{NL}, or else the nucleosome is extended indefinitely to create "delocalized" nucleosomal states (Yuan et al., 2005; Lee et al., 2007). The linker state can also be extended indefinitely with a probability P_{LL}, or else the transition is made back to the first nucleosome state N_1. Note that $P_{NL}+P_{DN}=1$, $P_{LL}+P_{LN}=1$. (B) Nucleosome positions prior to glucose addition at the *CHA1* promoter predicted by the HMM. Top: Log ratio of nucleosomal DNA to genomic DNA as determined by separate hybridizations to Affymetrix tiling arrays is plotted as a function of genomic position. Increasing values represent increasing protection from MNase digestion. Middle: Nucleosome positions predicted by the HMM (Zawadzki et al., 2009). Black trace represents predicted nucleosome occupancy, from unoccupied (0) to fully occupied (1). Blue trace represents the probability of starting a nucleosome at a given base pair. Bottom: Previously mapped *in vivo* nucleosome positions (Moreira and Holmberg, 1998) are shown as dark orange ovals. (See color plate 4).

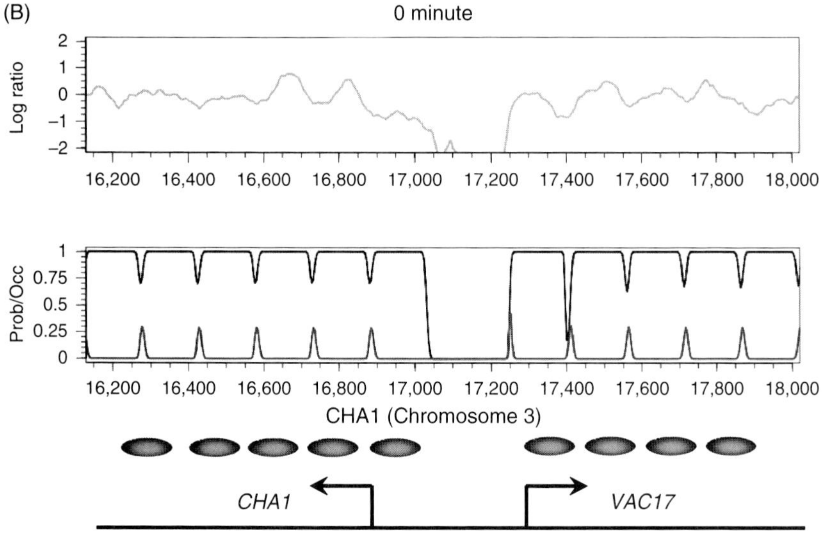

FIG. 11. (Continued)

two Gaussians. All fitting parameters were found by maximum likelihood using standard methods (Rabiner, 1989; Durbin et al., 1998). Note that the probability of starting a new nucleosome (P_{LN} in Fig. 11A), fit by maximum likelihood in Yuan et al. (2005) and Lee et al. (2007), was adjusted manually in Zawadzki et al. (2009) in order to achieve a pre-defined average nucleosome occupancy of ~80%. The manual adjustment reflects a lack of control over the zero-intensity baseline, which may shift depending on the relative amounts of DNA used in the nucleosomal and control samples.

In Lee et al. (2007) and Zawadzki et al. (2009), which employ high-density Affymetrix arrays, HMM parameters were fit for only a relatively small set of genomic loci (*e.g.* promoters and coding sequences of 19 genes whose expression was unaffected by adding glucose to the medium in Zawadzki et al. (2009)), and averaged. The resulting model was then run genome-wide with fixed parameters. As a typical example, an HMM prediction for the *CHA1* locus in which nucleosomes were previously mapped using low-throughput methods (Moreira and Holmberg, 1998) is shown in Fig. 11B.

Another important distinction between the three HMM approaches is that Yuan et al. (2005) and Lee et al. (2007) allow "delocalized" nucleosomes of arbitrary length (the modified topology of the transition matrix is shown in Fig. 11A with dashed lines), whereas Zawadzki et al. (2009) postulate only canonical, 148 bp nucleosomes and thus interpret longer stretches with high log-intensity ratios as shifted but overlapping nucleosome positions in distinct subpopulations of cells.

IV. Summary and Conclusions

It is amazing to see by how much the field of high-throughput mapping of nucleosome positions has advanced over just a few years. Whereas in 2005 the first tiling microarray covered only yeast chromosome III with 20-bp resolution (Yuan et al., 2005), today's high-throughput sequencing platforms can yield hundreds of millions of nucleosome positions. Microarray and high-throughput sequencing technologies are to some extent complementary: interrogating a genomic region with tiled microarrays yields nucleosome occupancy in that region, but the approach does not easily scale up to higher eukaryotes with much longer genomes. Thus the best currently available microarray data for bulk nucleosomes in *D. melanogaster* (122.6-Mbp genome length) is at 36-bp resolution (Mavrich et al., 2008b), while in *H. sapiens* (3,300-Mbp genome length) only select regions have been interrogated (Dennis et al., 2007; Ozsolak et al., 2007).

Unlike microarrays, high-throughput sequencing is not restricted to a particular genomic region: nucleosome sequence reads can come from any locus. However, even the latest parallel sequencing data sets do not provide enough read coverage to measure relative nucleosome occupancies in longer genomes, although the data can still be used to infer common nucleosome-positioning motifs (such as the periodic distribution of AA/AT/TA/TT dinucleotides shown in Fig. 5) and to study nucleosome organization in the vicinity of coding sequences and TF-binding sites. To circumvent the problem of low sequence read coverage, several recent studies of chromatin structure in human and fly chose to focus on the nucleosomes that incorporate the H2A.Z histone variant or have acetylated/methylated histone tails (Barski et al., 2007; Mavrich et al., 2008b). These partial maps yield comprehensive genome-wide coverage for a given nucleosomal subspecies but in general cannot be extrapolated to account for bulk nucleosomes.

Most collections of nucleosome sequences published to date have utilized single-end reads: only one end of the ~150-bp mononucleosome cores is sequenced and mapped onto the reference genome. The other end of the nucleosome core particle has to be inferred by adding 147 bp to the starting position of the sequence read. This procedure assumes that MNase treatment liberates nucleosome core particles precisely, without leaving undigested DNA at the nucleosome termini or digesting nucleosome-covered DNA. A measure of MNase-cutting precision is provided by plotting a correlation function between the starting coordinates of the sequence reads mapped onto the Watson (W) and Crick (C) DNA strands, respectively (Fig. 4). Although there is a distinct maximum at 0 bp lag (which corresponds to the 147-bp separation between the starting positions of the sequence reads mapped onto the W and C strands), the width of the peak clearly shows that the majority of mononucleosome cores are not isolated with a bp-level precision. The question of how precisely nucleosome core particles are located with respect to sequence read coordinates can be further addressed using paired-end reads (in which, as its name suggests, both ends of the DNA molecule are sequenced and mapped, so that its length is known exactly).

Another potential issue in nucleosome-positioning studies is MNase sequence specificity. It is unlikely that MNase actively repositions nucleosomes, because experiments done with and without nucleosome cross-linking yield similar patterns of nucleosome organization and similar sequence determinants of nucleosome positioning (Kaplan et al., 2009). However, the question of whether MNase-binding specificity can bias which nucleosomes get sequenced (by preferentially isolating mononucleosome cores flanked by "good" MNase-binding sites) is not fully resolved in the literature. While some authors argue that MNase-binding specificity is negligible (Ozsolak et al., 2007; Field et al., 2008), others explain observed A/T enrichment in regions flanking nucleosome core particles as a consequence of preferential cleavage by micrococcal nuclease (Johnson et al., 2006; Valouev et al., 2008).

High-throughput nucleosome-positioning studies have established a canonical picture of nucleosome organization in genic and intergenic regions. As shown in Fig. 3, most yeast genes are flanked by 5′ and 3′ NDRs, which help arrange genic nucleosomes into quasi-periodic arrays. Although most authors agree that intrinsic sequence preferences play a certain role in establishing *in vivo* nucleosome positions, there is a range of opinions as to how important this role is.

One series of studies argues on the basis of the strong correlation between sequence-based bioinformatics models and *in vivo* occupancy profiles that most nucleosomes in living cells are positioned by sequence (Segal et al., 2006; Field et al., 2008; Kaplan et al., 2009). A competing view, the so-called barrier model of nucleosome positioning, emphasizes that regular nucleosomal arrays can be created simply by steric exclusion (Mavrich et al., 2008a; Zhang et al., 2009). The ends of such arrays may be defined by nucleosome-excluding sequence elements, by "anchoring" nucleosomes optimized for binding affinity and/or stabilized by interactions with other proteins and protein complexes, or by DNA-bound TFs. According to this view, most nucleosomes need not be sequence-specific. It is interesting to note that there are significant differences between *in vitro* and *in vivo* nucleosome occupancy profiles: the 5′ NDR is significantly less pronounced *in vitro* and there are no nucleosome-size oscillations over ORFs (Fig. 3). Thus nucleosomes are not intrinsically ordered with respect to the TSS and borders of *in vivo* arrays are shaped by (yet unknown) external factors, for example, by interactions with the components of transcription initiation machinery (Zhang et al., 2009). Surprisingly, there is no difference between *in vitro* and *in vivo* 3′ NDRs, which thus appear to be established mainly through nucleosome-disfavoring sequences.

In this review we have deliberately not focused on the nucleosome occupancy in the vicinity of TF-binding sites. It is harder to see the general picture here because the relationship between nucleosomes and TFs is much more varied, and because precise genomic locations of TF-binding sites are not always known. Nevertheless, it appears that nucleosomes are intrinsically depleted over some types of TF-binding sites, whereas for other factors (such as Abf1 and Reb1 in yeast) there are marked differences between *in vivo* and *in vitro* nucleosome occupancies (Kaplan et al., 2009). In accordance with this view, chromatin appears surprisingly stable with respect to environmental (Shivaswamy et al., 2008; Zawadzki et al., 2009) and genetic (Whitehouse et al., 2007) perturbations. Overall, there are only minor changes in nucleosome positions and numbers, indicating that chromatin is largely "pre-set" for transcriptional response.

Computational approaches to predicting nucleosome occupancies can be based on either physics or bioinformatics. Physical models of nucleosome formation energies employ DNA elasticity theory (in some cases augmented with additional terms that take into account conformational entropy, intrinsic DNA curvature, etc.) to compute the sequence-dependent free energy of bending the 147-bp long DNA into a nucleosomal superhelix. Free energies computed at every base pair along the DNA sequence can then be used as input to the dynamic programming algorithm (Morozov et al., 2008), which solves the many-body problem of positioning multiple nucleosomes on DNA without steric overlap (even though most physics-based studies do not attempt to derive nucleosome occupancies from free-energy predictions). Unlike bioinformatics approaches, physical models do not utilize training sets of NPSs, which could lead to biased predictions if all sequences came from a particular genome or if the sequence set was not sufficiently large. However, elastic potentials typically depend on empirical coefficients such as equilibrium values of DNA geometric parameters. Such coefficients have to be estimated from structural data or molecular mechanics simulations, making the models dependent on the quality of the estimates.

In contrast to the bioinformatics approaches designed to search for DNA sequence signals that discriminate between nucleosome-enriched and nucleosome-depleted regions, physical models can explain observed sequence patterns in terms of elastic energies associated with DNA bending and the corresponding geometries of the nucleosomal DNA. If the DNA conformation is allowed to relax, such models are even capable of predicting the minimum-energy DNA conformation for comparison with crystal structures (Morozov et al., 2008). It appears that physical models can predict free energies and *in vitro* positions of single nucleosomes reconstituted on artificial and natural sequences reasonably well (Tolstorukov et al., 2007; Morozov et al., 2009; Scipioni et al., 2009). Their genome-wide accuracy is less clear: some papers do not make genome-wide predictions at all (Tolstorukov et al., 2007; Morozov et al., 2009) while others provide limited comparisons that do not include latest data sets or bioinformatics models (Miele et al., 2008; Scipioni et al., 2009). It would be of great interest to test physical models against each other and against their bioinformatics counterparts using latest high-throughput parallel sequencing data sets and a uniform set of performance metrics.

Bioinformatics approaches employ a wide range of statistical techniques (including support vector machines, wavelet analysis, and Markov models) to assign nucleosome-positioning scores. Although a rigorous comparison between all published models is not available, it is fair to say that the latest generation of bioinformatics models can be used to discriminate nucleosome-enriched from nucleosome-depleted regions with high accuracy and to predict general features of nucleosome organization such as 5′ and 3′ NDRs. An interesting hybrid approach has been developed by Eran Segal and coworkers (Segal et al., 2006; Field et al., 2008; Kaplan et al., 2009). As discussed in detail in Section III.B.1, the authors compute a nucleosome-positioning score by first making an alignment of a large number of nucleosomal sequences obtained from a high-throughput parallel sequencing run. Next they define a log-score based on: a) the dinucleotide distribution at each position in the nucleosomal site; and b) the position-independent distribution of 5 bp words inside and outside nucleosomes. The bioinformatics scores at each genomic position are then treated as "energies" and used to predict genome-wide nucleosome occupancy profiles by solving the many-body problem of positioning multiple nucleosomes on the genomic DNA (Section III.C). The latest iteration of this model is based on the *in vitro* nucleosome-positioning data set and is capable of predicting the *in vitro* occupancy profile with a correlation coefficient of 0.89 (Kaplan et al., 2009).

We expect that a more uniform view of the factors responsible for *in vivo* nucleosome positioning will emerge in the near future. Comparisons between various models and a careful analysis of the observed nucleosome sequence features will help establish the limits of applicability and the relative strengths and weaknesses of alternative modeling approaches. Future computational predictions of nucleosome positions should be equally applicable to genomic sequences from multiple organisms and to synthetic DNA, and will establish the relative importance of intrinsic nucleosome sequence preferences in maintaining and regulating *in vivo* chromatin.

Acknowledgments

The authors are grateful to George Locke for assistance with high-throughput sequencing data, and to Karl Zawadzki for carefully proofreading the manuscript. A.V.M. and D.T. were supported by a grant from the National Institutes of Health (R01 HG-004708). A.V.M. was also supported by an Alfred P. Sloan Research Fellowship.

References

Albert, I., Mavrich, T. N., Tomsho, L. P., Qi, J., Zanton, S. J., Schuster, S. C., et al., (2007). Translational and rotational settings of H2A.Z nucleosomes across the *Saccharomyces cerevisiae* genome. *Nature* **446**, 572–576.

Anderson, J. D., Widom, J., (2001). Poly(dA-dT) promoter elements increase the equilibrium accessibility of nucleosomal DNA target sites. *Mol. Cell. Biol.* **21**, 3830–3839.

Anselmi, C., Bocchinfuso, G., De Santis, P., Savino, M., Scipioni, A., (2000). A theoretical model for the prediction of sequence-dependent nucleosome thermodynamic stability. *Biophys. J.* **79**, 601–613.

Anselmi, C., De Santis, P., Paparcone, R., Savino, M., Scipioni, A., (2002). From the sequence to the superstructural properties of DNAs. *Biophys. Chem.* **95**, 23–47.

Bao, Y., White, C. L., Luger, K., (2006). Nucleosome core particles containing a poly (dA.dT) sequence element exhibit a locally distorted DNA structure. *J. Mol. Biol.* **361**, 617–624.

Barski, A., Cuddapah, S., Cui, K. R., Roh, T. Y., Schones, D. E., Wang, Z. B., et al., (2007). High-resolution profiling of histone methylations in the human genome. *Cell* **129**, 823–837.

Bernstein, B. E., Liu, C. L., Humphrey, E. L., Perlstein, E. O., Schreiber, S. L., (2004). Global nucleosome occupancy in yeast. *Genome Biol.* **5**, R62.

Boeger, H., Griesenbeck, J., Strattan, J. S., Kornberg, R. D., (2003). Nucleosomes unfold completely at a transcriptionally active promoter. *Mol. Cell.* **11**, 1587–1598.

Chakravarthy, S., Gundimella, S. K., Caron, C., Perche, P. Y., Pehrson, J. R., Khochbin, S., et al., (2005). Structural characterization of the histone variant macroH2A. *Mol. Cell. Biol.* **25**, 7616–7624.

Chodaparambil, J. V., Barbera, A. J., Lu, X., Kaye, K. M., Hansen, J. C., Luger, K., (2007). A charged and contoured surface on the nucleosome regulates chromatin compaction. *Nat. Struct. Mol. Biol.* **14**, 1105–1107.

Cohanim, A. B., Kashi, Y., Trifonov, E. N., (2006). Three sequence rules for chromatin. *J. Biomol. Struct. Dyn.* **23**, 559–566.

Dennis, J. H., Fan, H. Y., Reynolds, S. M., Yuan, G. C., Meldrim, J. C., Richter, D. J., et al., (2007). Independent and complementary methods for large-scale structural analysis of mammalian chromatin. *Genome Res.* **17**, 928–939.

De Santis, P., Palleschi, A., Morosetti, S., Savino, M., (1986). Structures and Superstructures in Periodical Polynucleotides. Pergamon Press, New York.

Dohmen, R. J., Varshavsky, A., (2005). Heat-inducible degron and the making of conditional mutants. *Methods Enzymol.* **399**, 799–822.

Dorigo, B., Schalch, T., Kulangara, A., Duda, S., Schroeder, R. R., Richmond, T. J., (2004). Nucleosome arrays reveal the two-start organization of the chromatin fiber. *Science* **306**, 1571–1573.

Draker, R., Cheung, P., (2009). Transcriptional and epigenetic functions of histone variant H2A.Z. *Biochem. Cell Biol.* **87**, 19–25.

Durbin, R., Eddy, S., Krogh, A., Mitchison, G., (1998). Biological Sequence Analysis: Probabilistic Models of Proteins and Nucleic Acids. Cambridge University Press, Cambridge, UK New York.

Edayathumangalam, R. S., Weyermann, P., Gottesfeld, J. M., Dervan, P. B., Luger, K., (2004). Molecular recognition of the nucleosomal "supergroove". *Proc. Natl. Acad. Sci. USA* **101**, 6864–6869.

Fedor, M. J., Lue, N. F., Kornberg, R. D., (1988). Statistical positioning of nucleosomes by specific protein-binding to an upstream activating sequence in yeast. *J. Mol. Biol.* **204**, 109–127.

Felsenfeld, G., Groudine, M., (2003). Controlling the double helix. *Nature* **421**, 448–453.

Field, Y., Kaplan, N., Fondufe-Mittendorf, Y., Moore, I. K., Sharon, E., Lubling, Y., et al., (2008). Distinct Modes of Regulation by Chromatin Encoded through Nucleosome Positioning Signals. *PLoS Comput. Biol.* **4**, e1000216(25).

Fragoso, G., John, S., Roberts, M. S., Hager, G. L., (1995). Nucleosome Positioning on the Mmtv Ltr Results from the Frequency-Biased Occupancy of Multiple Frames. *Genes Dev.* **9**, 1933–1947.

Goodsell, D. S., Dickerson, R. E., (1994). Bending and curvature calculations in B-DNA. *Nucleic Acids Res.* **22**, 5497–5503.

Hartley, P. D., Madhani, H. D., (2009). Mechanisms that Specify Promoter Nucleosome Location and Identity. *Cell* **137**, 445–458.

Horz, W., Altenburger, W., (1981). Sequence specific cleavage of DNA by micrococcal nuclease. *Nucleic Acids Res.* **9**, 2643–2658.

Ioshikhes, I. P., Albert, I., Zanton, S. J., Pugh, B. F., (2006). Nucleosome positions predicted through comparative genomics. *Nat. Genet.* **38**, 1210–1215.

Ioshikhes, I., Bolshoy, A., Derenshteyn, K., Borodovsky, M., Trifonov, E. N., (1996). Nucleosome DNA sequence pattern revealed by multiple alignment of experimentally mapped sequences. *J. Mol. Biol.* **262**, 129–139.

Ito, T., Bulger, M., Pazin, M. J., Kobayashi, R., Kadonaga, J. T., (1997). ACF, an Isw1-containing and ATP-utilizing chromatin assembly and remodeling factor. *Cell* **90**, 145–155.

Iyer, V., Struhl, K., (1995). Poly(dA:dT), a ubiquitous promoter element that stimulates transcription via its intrinsic DNA structure. *EMBO J.* **14**, 2570–2579.

Jiang, C. Z., Pugh, B. F., (2009). Nucleosome positioning and gene regulation: advances through genomics. *Nat. Rev. Genet.* **10**, 161–172.

Johnson, S. M., Tan, F. J., McCullough, H. L., Riordan, D. P., Fire, A. Z., (2006). Flexibility and constraint in the nucleosome core landscape of *Caenorhabditis elegans* chromatin. *Genome Res.* **16**, 1505–1516.

Kaplan, N., Moore, I. K., Fondufe-Mittendorf, Y., Gossett, A. J., Tillo, D., Field, Y., et al., (2009). The DNA-encoded nucleosome organization of a eukaryotic genome. *Nature* **458**, 362–366.

Kato, M., Onishi, Y., Wada-Kiyama, Y., Abe, T., Ikemura, T., Kogan, S., et al., (2003). Dinucleosome DNA of human K562 cells: experimental and computational characterizations. *J. Mol. Biol.* **332**, 111–125.

Khorasanizadeh, S., (2004). The nucleosome: from genomic organization to genomic regulation. *Cell* **116**, 259–272.

Kornberg, R. D., Stryer, L., (1988). Statistical distributions of nucleosomes: nonrandom locations by a stochastic mechanism. *Nucleic Acids Res.* **16**, 6677–6690.

Kornberg, R. D., Thomas, J. O., (1974). Chromatin structure; oligomers of the histones. *Science* **184**, 865–868.

Lee, C. K., Shibata, Y., Rao, B., Strahl, B. D., Lieb, J. D., (2004). Evidence for nucleosome depletion at active regulatory regions genome-wide. *Nat. Genet.* **36**, 900–905.

Lee, W., Tillo, D., Bray, N., Morse, R. H., Davis, R. W., Hughes, T. R., et al., (2007). A high-resolution atlas of nucleosome occupancy in yeast. *Nat. Genet.* **39**, 1235–1244.

Lowary, P. T., Widom, J., (1997). Nucleosome packaging and nucleosome positioning of genomic DNA. *Proc. Natl. Acad. Sci. USA* **94**, 1183–1188.

Lu, X. J., El Hassan, M. A., Hunter, C. A., (1997a). Structure and conformation of helical nucleic acids: analysis program (SCHNAaP). *J. Mol. Biol.* **273**, 668–680.

Lu, X. J., El Hassan, M. A., Hunter, C. A., (1997b). Structure and conformation of helical nucleic acids: rebuilding program (SCHNArP). *J. Mol. Biol.* **273**, 681–691.

Lubliner, S., Segal, E., (2009). Modeling interactions between adjacent nucleosomes improves genome-wide predictions of nucleosome occupancy. *Bioinformatics* **25**, I348–I355.

Luger, K., Mader, A. W., Richmond, R. K., Sargent, D. F., Richmond, T. J., (1997). Crystal structure of the nucleosome core particle at 2.8 Å resolution. *Nature* **389**, 251–260.

Mallat, S. G., (1999). A Wavelet Tour of Signal Processing, second ed., Academic, San Diego, Calif.

Mavrich, T. N., Ioshikhes, I. P., Venters, B. J., Jiang, C., Tomsho, L. P., Qi, J., et al., (2008a). A barrier nucleosome model for statistical positioning of nucleosomes throughout the yeast genome. *Genome Res.* **18**, 1073–1083.

Mavrich, T. N., Jiang, C. Z., Ioshikhes, I. P., Li, X. Y., Venters, B. J., Zanton, S. J., et al., (2008b). Nucleosome organization in the Drosophila genome. *Nature* **453**, 358– 362.

Miele, V., Vaillant, C., d'Aubenton-Carafa, Y., Thermes, C., Grange, T., (2008). DNA physical properties determine nucleosome occupancy from yeast to fly. *Nucleic Acids Res.* **36**, 3746–3756.

Mito, Y., Henikoff, J. G., Henikoff, S., (2007). Histone replacement marks the boundaries of cis-regulatory domains. *Science* **315**, 1408–1411.

Moreira, J. M.A., Holmberg, S., (1998). Nucleosome structure of the yeast CHA1 promoter: analysis of activation-dependent chromatin remodeling of an RNA-polymerase-II-transcribed gene in TBP and RNA pol II mutants defective *in vivo* in response to acidic activators. *EMBO J.* **17**, 6028–6038.

Morozov, A.V., Fortney, K., Gaykalova, D.A., Studitsky, V.M., Widom, J., Siggia, E.D. (2008) Extrinsic and Intrinsic Nucleosome Positioning Signals. www.arXiv.org:0805.4017. (accessed 07.01.2009)

Morozov, A. V., Fortney, K., Gaykalova, D. A., Studitsky, V. M., Widom, J., Siggia, E. D., (2009). Using DNA mechanics to predict *in vitro* nucleosome positions and formation energies. *Nucleic Acids Res.* **37**, 4707–4722.

Morse, R. H., (2007). Transcription factor access to promoter elements. *J. Cell. Biochem.* **102**, 560–570.

Muthurajan, U. M., Bao, Y., Forsberg, L. J., Edayathumangalam, R. S., Dyer, P. N., White, C. L., et al., (2004). Crystal structures of histone Sin mutant nucleosomes reveal altered protein–DNA interactions. *EMBO J.* **23**, 260–271.

Nagalakshmi, U., Wang, Z., Waern, K., Shou, C., Raha, D., Gerstein, M., et al., (2008). The transcriptional landscape of the yeast genome defined by RNA sequencing. *Science* **320**, 1344–1349.

Olson, W. K., Gorin, A. A., Lu, X. J., Hock, L. M., Zhurkin, V. B., (1998). DNA sequence-dependent deformability deduced from protein–DNA crystal complexes. *Proc. Natl. Acad. Sci. USA* **95**, 11163–11168.

Ozsolak, F., Song, J. S., Liu, X. S., Fisher, D. E., (2007). High-throughput mapping of the chromatin structure of human promoters. *Nat. Biotechnol.* **25**, 244–248.

Pazin, M. J., Bhargava, P., Geiduschek, E. P., Kadonaga, J. T., (1997). Nucleosome mobility and the maintenance of nucleosome positioning. *Science* **276**, 809–812.

Peckham, H. E., Thurman, R. E., Fu, Y. T., Stamatoyannopoulos, J. A., Noble, W. S., Struhl, K., et al., (2007). Nucleosome positioning signals in genomic DNA. *Genome Res.* **17**, 1170–1177.

Percus, J. K., (1976). Equilibrium State of a Classical Fluid of Hard Rods in an External Field. *J. Stat. Phys.* **15**, 505–511.

Rabiner, L. R., (1989). A Tutorial on Hidden Markov-Models and Selected Applications in Speech Recognition. *Proc. IEEE.* **77**, 257–286.

Richardfoy, H., Hager, G. L., (1987). Sequence-Specific Positioning of Nucleosomes over the Steroid-Inducible Mmtv Promoter. *EMBO J.* **6**, 2321–2328.

Richmond, T. J., Davey, C. A., (2003). The structure of DNA in the nucleosome core. *Nature* **423**, 145–150.

Roth, S. Y., Dean, A., Simpson, R. T., (1990). Yeast alpha 2 repressor positions nucleosomes in TRP1/ARS1 chromatin. *Mol. Cell. Biol.* **10**, 2247–2260.

Satchwell, S. C., Drew, H. R., Travers, A. A., (1986). Sequence periodicities in chicken nucleosome core DNA. *J. Mol. Biol.* **191**, 659–675.

Schalch, T., Duda, S., Sargent, D. F., Richmond, T. J., (2005). X-ray structure of a tetranucleosome and its implications for the chromatin fibre. *Nature* **436**, 138–141.

Schones, D. E., Cui, K. R., Cuddapah, S., Roh, T. Y., Barski, A., Wang, Z. B., et al., (2008). Dynamic regulation of nucleosome positioning in the human genome. *Cell* **132**, 887–898.

Schwab, D. J., Bruinsma, R. F., Rudnick, J., Widom, J., (2008). Nucleosome switches. *Phys. Rev. Lett.* **100**, 228105.

Scipioni, A., Morosetti, S., De Santis, P., (2009). A statistical thermodynamic approach for predicting the sequence-dependent nucleosome positioning along genomes. *Biopolymers* **91**, 1143–1153.

Segal, E., Fondufe-Mittendorf, Y., Chen, L. Y., Thastrom, A., Field, Y., Moore, I. K., et al., (2006). A genomic code for nucleosome positioning. *Nature* **442**, 772–778.

Segal, E., Widom, J., (2009). Poly(dA:dT) tracts: major determinants of nucleosome organization. *Curr. Opin. Struct. Biol.* **19**, 65–71.

Shivaswamy, S., Bhinge, A., Zhao, Y. J., Jones, S., Hirst, M., Iyer, V. R., (2008). Dynamic remodeling of individual nucleosomes across a eukaryotic genome in response to transcriptional perturbation. *PLoS Biol.* **6**, 618–630.

Shrader, T. E., Crothers, D. M., (1989). Artificial nucleosome positioning sequences. *Proc. Natl. Acad. Sci. USA* **86**, 7418–7422.

Shrader, T. E., Crothers, D. M., (1990). Effects of DNA-sequence and histone–histone interactions on nucleosome placement. *J. Mol. Biol.* **216**, 69–84.

Stormo, G. D., Fields, D. S., (1998). Specificity, free energy and information content in protein–DNA interactions. *Trends Biochem. Sci.* **23**, 109–113.

Strahl, B. D., Allis, C. D., (2000). The language of covalent histone modifications. *Nature* **403**, 41–45.

Suter, B., Schnappauf, G., Thoma, F., (2000). Poly(dA.dT) sequences exist as rigid DNA structures in nucleosome-free yeast promoters *in vivo*. *Nucleic Acids Res.* **28**, 4083–4089.

Suto, R. K., Clarkson, M. J., Tremethick, D. J., Luger, K., (2000). Crystal structure of a nucleosome core particle containing the variant histone H2A.Z. *Nat. Struct. Biol.* **7**, 1121–1124.

Suto, R. K., Edayathumangalam, R. S., White, C. L., Melander, C., Gottesfeld, J. M., Dervan, P. B., et al., (2003). Crystal structures of nucleosome core particles in complex with minor groove DNA-binding ligands. *J. Mol. Biol.* **326**, 371–380.

Thastrom, A., Bingham, L. M., Widom, J., (2004). Nucleosomal locations of dominant DNA sequence motifs for histone–DNA interactions and nucleosome positioning. *J. Mol. Biol.* **338**, 695–709.

Thastrom, A., Lowary, P. T., Widlund, H. R., Cao, H., Kubista, M., Widom, J., (1999). Sequence motifs and free energies of selected natural and non-natural nucleosome positioning DNA sequences. *J. Mol. Biol.* **288**, 213–229.

Tolstorukov, M. Y., Colasanti, A. V., McCandlish, D. M., Olson, W. K., Zhurkin, V. B., (2007). A novel roll-and-slide mechanism of DNA folding in chromatin: implications for nucleosome positioning. *J. Mol. Biol.* **371**, 725–738.

Vaillant, C., Audit, B., Arneodo, A., (2007). Experiments confirm the influence of genome long-range correlations on nucleosome positioning. *Phys. Rev. Lett.* **99**, 218103.

Valouev, A., Ichikawa, J., Tonthat, T., Stuart, J., Ranade, S., Peckham, H., et al., (2008). A high-resolution, nucleosome position map of *C. elegans* reveals a lack of universal sequence-dictated positioning. *Genome Res.* **18**, 1051–1063.

Vapnik, V. N., (1998). Statistical Learning Theory. Wiley, New York.

Wallrath, L. L., Lu, Q., Granok, H., Elgin, S. C.R., (1994). Architectural variations of inducible eukaryotic promoters – preset and remodeling chromatin structures. *Bioessays* **16**, 165–170.

Wang, J. P., Fondufe-Mittendorf, Y., Xi, L., Tsai, G. F., Segal, E., Widom, J., (2008). Preferentially quantized linker DNA lengths in *Saccharomyces cerevisiae*. *PLoS Comput. Biol.* **4**, e1000175.

Whitehouse, I., Rando, O. J., Delrow, J., Tsukiyama, T., (2007). Chromatin remodelling at promoters suppresses antisense transcription. *Nature* **450**, 1031–1035.

Widom, J., (1992). A relationship between the helical twist of DNA and the ordered positioning of nucleosomes in all eukaryotic cells. *Proc. Natl. Acad. Sci. USA* **89**, 1095–1099.

Wingert, L., Von Hippel, P. H., (1968). The conformation dependent hydrolysis of DNA by micrococcal nuclease. *Biochim. Biophys. Acta* **157**, 114–126.

Yuan, G. C., Liu, J. S., (2008). Genomic sequence is highly predictive of local nucleosome depletion. *PLoS Comput. Biol.* **4**, 0164–0174.

Yuan, G. C., Liu, Y. J., Dion, M. F., Slack, M. D., Wu, L. F., Altschuler, S. J., et al., (2005). Genome-scale identification of nucleosome positions in *S. cerevisiae*. *Science* **309**, 626–630.

Zaman, S., Lippman, S. I., Schneper, L., Slonim, N., Broach, J. R., (2009). Glucose regulates transcription in yeast through a network of signaling pathways. *Mol. Syst. Biol.* **5**, 245.

Zawadzki, K. A., Morozov, A. V., Broach, J. R., (2009). Chromatin-dependent Transcription factor accessibility rather than nucleosome remodeling predominates during global transcriptional restructuring in *Saccharomyces cerevisiae*. *Mol. Biol. Cell.* **20**, 3503–3513.

Zhang, Y., Moqtaderi, Z., Rattner, B. P., Euskirchen, G., Snyder, M., Kadonaga, J. T., et al., (2009). Intrinsic histone-DNA interactions are not the major determinant of nucleosome positions *in vivo*. *Nat. Struct. Mol. Biol.* **16**, 847–852.

PROTEIN HOMOREPEATS: SEQUENCES, STRUCTURES, EVOLUTION, AND FUNCTIONS

By JULIEN JORDA AND ANDREY V. KAJAVA

Centre de Recherches de Biochimie Macromoléculaire UMR 5237, CNRS, University of Montpellier 1 and 2, Montpellier, France

Abstract

The vast majority of protein sequences are aperiodic; they do not have any strong bias in the amino acid composition, and they use a subtle mixture of all or most of the 20 amino acid residues to code a great

ADVANCES IN PROTEIN CHEMISTRY AND STRUCTURAL BIOLOGY, Vol. 79
DOI: 10.1016/S1876-1623(10)79002-7

number of various structures and functions. In this context, homorepeats, runs of a single amino acid residue, represent unusual, eye-catching motifs in proteins. Despite the sequence simplicity and relatively small size, the homorepeat runs have a strong potential for molecular interactions due to the excessively high local concentration of a certain physico-chemical property. Appearance of such runs within proteins may give them new structural and functional features. An increasing number of studies demonstrate the abundance of these motifs in proteins, their important roles in biological processes, and their link to a number of hereditary and age-related diseases. In this chapter, we summarize data on the distribution of homorepeats in proteomes and on their structural properties, evolution, and functions.

I. Introduction

Amino acid homorepeats, also called homopeptide repeats, single amino acid repeats, or amino acid runs, represent protein regions with reiterations of a single residue. The homorepeats look unusually simple compared to typical protein sequences which are characterized by high complexity and an absence of strong bias in the amino acid composition.

A series of studies suggests that these regions represent the protein equivalent of junk DNA sequences that mostly serve as linkers between functionally important domains (Huntley and Golding, 2002; Newfeld et al., 1994; Gatti et al., 1994; Sumiyama et al., 1996). However, it is becoming apparent that homorepeats play crucial roles in key biological processes. They tend to belong to specific functional classes of proteins such as transcription factors, protein kinases, and proteins of ion storage and exchange (Karlin and Burge, 1996). Variation of homorepeat length which modulate transcriptional activation can be the major source of phenotypic diversity (Gerber et al., 1994; Fondon and Garner, 2004). The runs of poly-A and poly-G (here and in the rest of the chapter the one-letter amino acid code is used) are indispensable to provide necessary mechanical properties to a number of extracellular structural proteins (Fraser and MacRae, 1973; Altman et al., 2003)

Homorepeat regions are also attracting considerable attention because of their links to a number of hereditary and age-related diseases. In particular, several poly-Q-containing proteins are associated with inherited neurodegenerative disorders, including Huntington's disease, spinobulbar muscular atrophy, dentatorubral-pallidoluysian atrophy, and different forms of spinocerebellar ataxia (Orr and Zoghbi, 2007). Expansion of poly-A in polyadenine-binding protein 2 is associated with oculopharyngeal muscular dystrophy (Brais et al., 1998). Several other genes have been identified in which poly-A expansions may cause human diseases, including the Artistaless-related homeobox protein in X-linked mental retardation and epilepsy (Stromme et al., 2002) and the SRY-box 3 in X-linked mental retardation (Laumonnier et al., 2002). A growing number of studies suggest that homorepeats may have a broader role in human diseases than was previously recognized (Siwach and Ganesh, 2008). In accordance with the Online Mendelian Inheritance in Man (OMIM) annotation (Hamosh et al., 2002), more than 40% of proteins containing multiple runs of homorepeats are associated with different human diseases (Karlin et al., 2002).

Another unusual property of the homorepeat regions is their exceptionally high length polymorphism. This polymorphism is observed within populations of the same organisms, between different cell types within an organism, and even depends on the age of an individual (Ellegren, 2000; Brouwer et al., 2009). The growing attention to homorepeats reflects both the discovery of homorepeat instability and a high capacity for cytotoxicity of these regions. It is still unknown why potentially harmful and genetically unstable homorepeats have been maintained throughout evolution.

In living organisms, homopeptides can also be of nonribosomal origin. Over the last few years, a large family of enzymes have been characterized whose members catalyze attachment of multiple amino acids to a γ-carboxyl group of a glutamate residue of target proteins (Edde et al., 1990; Redeker et al., 1994; Janke et al., 2005).

Unusual structural properties of homorepeats, including their ability to form crystalline structures, and to impart high concentrations of the same physico-chemical properties (hydrophobicity, charge, rigidity, etc.) within a given region, create an opportunity for successful application of homorepeat-containing molecules in biotechnology and materials science. Known examples include the poly-H tag, which is widely used to purify recombinant proteins, poly-K coating of culture dishes, which provides the properties required for cell adhesion and proliferation (Ai et al., 2003), and poly-G molecules, which can self-assemble into different supramolecular structures (Kajava, 1999; Bovin et al., 2008).

The present survey was inspired by the fact that genome-sequencing projects are revealing a large number of protein sequences, and this information base has made it possible to conduct a detailed bioinformatics analysis of the homorepeat runs (Singh et al., 2004; Faux et al., 2005). In this chapter we use a protein repeat database (PRDB) that is accessible at http://bioinfo.montp.cnrs.fr/?r=repeatDB (Jorda and Kajava, 2009) to confirm or supplement the previous observations and to generate new conclusions. This bioinformatics tool can be used to analyze amino acid compositions, length distributions, cellular localization, functions, and other properties of the homorepeats.

II. Distribution of Homorepeats in Proteomes

A. *What Is the Minimal Homorepeat Run That Should Be Taken into Consideration?*

The exponential growth of genomic data has led to increasing efforts to analyze the distribution of homorepeats in coding regions of the genomes (Marcotte et al., 1999; Katti et al., 2001; Karlin et al., 2002; Faux et al., 2005). One of the first questions is "What is the minimum length of the homorepeat runs that should be taken into consideration?" The frequency of repetitive regions in proteins increases exponentially with the decrease of their length (Marcotte et al., 1999); therefore, the choice of the minimum length of the run can affect significantly the size and content of the analyzed set of homorepeats. The shorter the run, the less impact it may have on the structure and function of the protein. Very short runs may occur simply by chance, and not because of their functionality. Depending on the statistical models and assumptions, the minimum length used in the genome analyses is 5–7 residues (Karlin, 1995; Katti et al., 2001; Karlin et al., 2002; Faux et al., 2005). These values, based on standard random models of sequence evolution, may not be the most accurate in view of recent data that the frequency distribution of genomic motifs may arise from duplicative evolutionary processes, and not necessarily from selection based on point mutations and deletions

(Csuros et al., 2007). Nevertheless, the calculated minimum length of 5–7 residues is close to the limit at which some of the homorepeats start to affect structure and function. For example, proteins containing alanine runs of more than 10 residues demonstrate an enhanced tendency to aggregate (Fan et al., 2001). Furthermore, fusion of six histidines with a protein is sufficient for efficient purification of this recombinant protein via binding of the histidine tag to nickel-containing columns (Hengen, 1995).

B. Homorepeats Are Predominantly Hydrophilic

Several general conclusions became apparent across all the data. For example, homorepeats from all kingdoms are predominantly hydrophilic or composed of small residues such as Gly, Pro, and Ala, while large aliphatic, aromatic residues and Cys are rare (Green and Wang, 1994; Katti et al., 2001; Karlin et al., 2002; Faux et al., 2005) (Fig. 1). Furthermore, if runs of a hydrophobic residue do occur, their length tends to be less than 15 residues, while some hydrophilic runs can be longer than 50 residues (Karlin et al., 2002; Faux et al., 2005).

The most frequent runs made of large hydrophobic residues contain leucine (Fig. 1). A majority of poly-L runs (e.g., 90% in human proteins) occurs within the first 40 amino acid residues of the amino terminus identified as a part of signal-peptide sequences (Karlin et al., 2002). Other homorepeats of large hydrophobic residues do not have preferred locations within protein sequences. The great bulk of these hydrophobic runs represent transmembrane regions of cell membrane proteins.

The large number of hydrophilic runs can be explained by their lower propensity to aggregate in comparison to the hydrophobic stretches. Interestingly, runs of small nonpolar alanine are very frequent, despite their hydrophobic properties, the ability to form transmembrane regions, and their link to cytotoxicity (Fan et al., 2001). The abundance of poly-A may be associated with the small size of the side chain that attenuates the aggregation potential of the homorepeat thereby allowing a wider range of functions, including membrane anchoring, binding to proteins and DNA, and contributing to the rigidity of extracellular tissues such as elastins.

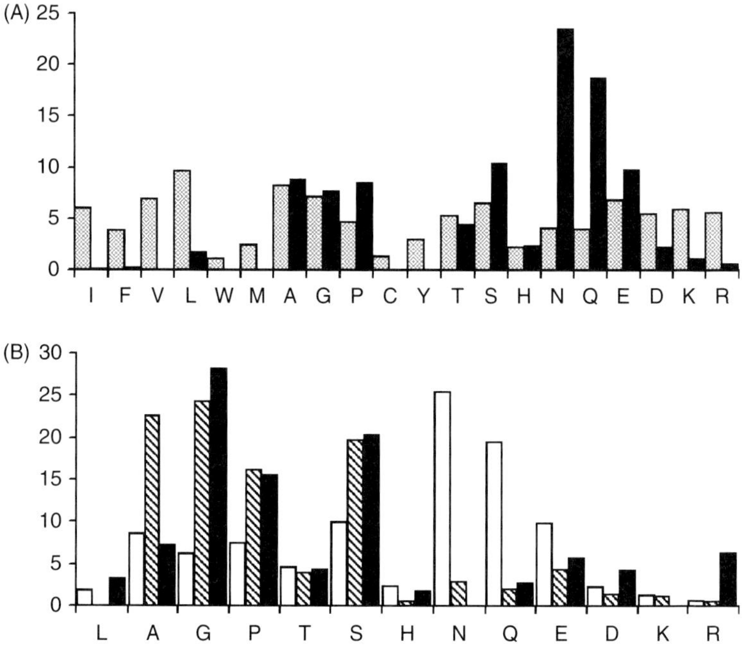

FIG. 1. Amino acid frequencies (%) ranged in order of decreasing hydrophobicity (Eisenberg et al., 1984). (A) Black and gray bars represent homorepeats ($\geq$7 residues) and all protein sequences, correspondingly, from Swiss-Prot (release July, 2009). The homorepeat data used in this and the other figures were taken from the PRDB. (B) Open bars represent homorepeats from eukaryotic proteins, the hatched bars show homorepeats from bacterial proteins, and the black bars correspond to viral proteins of Swiss-Prot. Poly-I, poly-F, poly-V, poly-W, poly-M, and poly-C homorepeats are omitted due to their low occurrences in all tested organisms. The remaining amino acid residues are arranged in accordance with the hydrophobic scale.

C. Homorepeats Are Much More Prevalent in Eukaryotic Than in Viral or Prokaryotic Proteins

Another conclusion revealed from numerous studies is that eukaryotes show a much higher frequency of homorepeats compared to prokaryotes (Karlin and Burge, 1996; Marcotte et al., 1999; Huntley and Golding, 2000; Faux et al., 2005). The analysis of Swiss-Prot homorepeats from PRDB shows that on an average 4.5% of the eukaryotic proteins have at least one homorepeat run (Fig. 2). Among the noneukaryotic organisms,

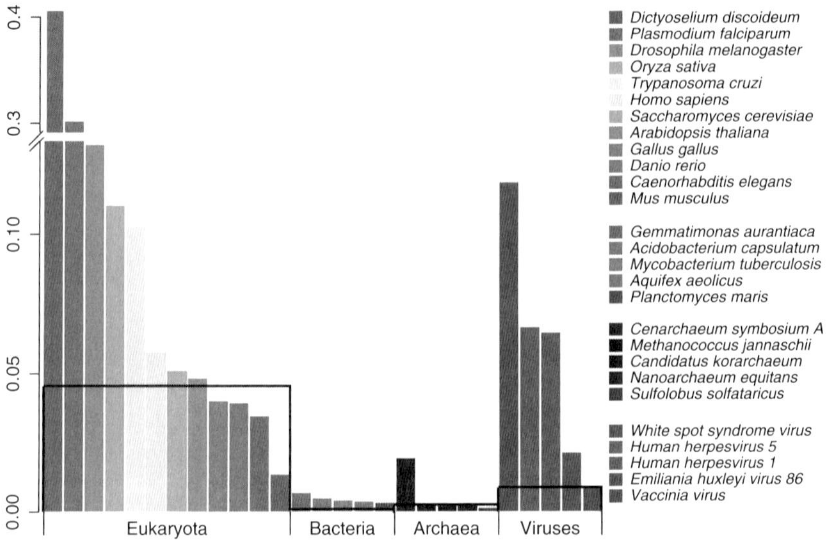

FIG. 2. Frequencies of homorepeat-containing proteins (HCP) in different proteomes. A representative set of 27 proteomes was clustered according to taxonomic domains (eukaryota, bacteria, archaea, and viruses). They have been gathered from the RefSeq database of NCBI. Within each taxonomic group, the proteomes are arranged in order of decreasing HCP frequencies. Open bars show the HCP frequencies in Swiss-Prot databank grouped by their taxonomic domains. Within the taxonomic groups, many HCP frequencies of individual proteomes are higher than the average Swiss-Prot frequency. This can be explained by the fact that our selection of the proteomes was biased toward organisms with a high number of homorepeats.

viruses have the highest frequency of homorepeat-containing proteins (about 1%) followed by archaea (0.2%) and bacteria (0.06%).

In eukaryotes, the most common homorepeats (in order of increasing frequency) are poly-Q, poly-N, poly-A, poly-S, poly-G, and poly-P. In prokaryotes, they are poly-S, poly-G, poly-A, and poly-P, and in viruses, they are poly-G, poly-S, poly-P, and poly-A (Fig. 1). A noticeable disparity is the rare occurrence of poly-Q and poly-N in prokaryotes and viruses that are the most common homorepeats in eukaryotes.

The acidic runs (poly-E and poly-D) exceed the basic runs (poly-K and poly-R) by a factor of 2 to 4 in eukaryotes while in bacterial genomes the proportions of the charged runs are almost equal (Karlin et al., 2002).

According to PRDB data, the proportion of basic runs is similar in eukaryotes and prokaryotes (about 6–7%) (Fig. 1). This indicates that the disproportion of the charged runs in eukaryotes is due to an increase of acidic runs. The acidic runs are not only more numerous in eukaryotes, but also far longer than the basic runs. The longest acidic runs are about 40–70 residues in contrast to the longest basic runs of 20–25 residues. Our analysis shows that a set of viral homorepeats from PRDB displays a prevalence of negatively charged runs compared to positively charged runs in a way similar to eukaryotes (Fig. 1). Surprisingly, in viruses, positively charged homorepeats are represented only by poly-R. The reason for the absence of poly-K is unclear.

D. *Frequencies and Types of Homorepeats Vary Greatly between Eukaryotes*

The higher frequency of homorepeats in eukaryotic proteins compared to prokaryotic ones is in an agreement with the previous conclusion regarding an increase of the homorepeat frequency with the genome size (Hancock and Simon, 2005). However, subsequent analysis of several complete genomes revealed that this correlation is not observed in either prokaryotic or eukaryotic proteomes (Huntley and Golding, 2000; Karlin et al., 2002; Singh et al., 2004; Faux et al., 2005). Table I shows how the percentages of proteins with at least one run of homorepeats in eukaryotes vary in order of the increase of the genome size. This comparison indicates a striking difference between the organisms. *Plasmodium falciparum* and *Dictyostelium discoideum* have

TABLE I
Genome size and percentage of homorepeat-containing proteins in different eukaryotes

Organism	Homorepeats (%)[a]	Size of genome (Mb)	Number of proteins
Saccharomyces cerevisiae	5	13	5,884
Plasmodium falciparum	30	23	5,271
Plasmodium yoelii	9.4	23	7,353
Caenorhabditis elegans	3.4	100	23,906
Arabidopsis thaliana	4.8	140	33,402
Drosophila melanogaster	13.2	165	21,099
Homo sapiens	5.7	3,300	37,392
Dictyostelium discoideum	40.4	34	13,408

[a] Percentage of homorepeats from PRDB.

relatively small genomes in the listed organisms but encode the highest numbers of proteins with homorepeats (36% and 40.4% of proteins, correspondingly). When grouping organisms differently, for example, considering only vertebrates, the variability is also high. The homorepeats are 10 times more frequent in mammals than in amphibians, birds, or fishes. Surprisingly, a noticeable difference is also observed within the same family of malaria parasites. The rodent parasite *Plasmodium yoelii* contains almost three times less proteins with homorepeats than human parasite *P. falciparum* (Singh et al., 2004). Currently there is no plausible explanation for this variation. Comparison of homorepeats within either bacterial, archaeal, or viral genomes also shows strong variations of the repeat frequencies (Fig. 2), and this variation does not correlate with the size of the genomes.

The other remarkable difference observed in homorepeats of eukaryotes is organism-specific diversity in the proportions of the types of the amino acid runs (Karlin et al., 2002; Faux et al., 2005). Although poly-Q, poly-N, poly-A, poly-S, poly-G, and poly-P runs account for a significant proportion of the runs in almost all eukaryotes, human proteins are most abundant (in increasing order) in poly-Q, poly-P, poly-L, and poly-A runs. Fly proteins are most abundant in poly-Q, poly-A, poly-S, and poly-G, *Caenorhabditis elegans* proteins in poly-S, poly-T, poly-P, and poly-Q, proteins from *Saccharomyces cerevisiae* in poly-S, poly-Q, poly-E, poly-D, and poly-N, *Arabidopsis thaliana* proteins in poly-S, poly-E, poly-G, and poly-P, and *P. falciparum* proteins in poly-S, poly-T, and poly-N. *Drosophila melanogaster* is enriched in poly-Q repeats (3.5 times more than *Homo sapiens*) but poly-Q runs are absent in *P. falciparum* (Faux et al., 2005). The most striking observation is that, for unknown reasons, poly-N runs are extremely rare in vertebrates while that of the chemically and structurally similar poly-Q are very frequent. In contrast, in nonvertebrates, poly-N represents an important fraction of homorepeats but not poly-Q (Kreil and Kreil, 2000; Karlin et al., 2002; Faux et al., 2005).

E. Distribution and Localization of Homorepeats within Proteins

The occurrence of more than one run in a given protein (called multiple homorepeat runs) is quite frequent, especially in eukaryotes. For example, about 23% of all homorepeat-containing proteins have more than one run including 24% of these types of proteins in eukaryotes and only 9% for prokaryotes (Faux et al., 2005). In most cases the multiple runs in a given protein consist of the same amino acid residue. Among the pairs of different types of homorepeats in one protein, the most frequent pairs include either poly-N or poly-Q runs, which also individually represent the most frequent homorepeats. Strong association is observed for the pairs poly-A and poly-Q, poly-H and poly-Q, poly-D and poly-N, poly-K and poly-N, poly-F and poly-N, and poly-T and poly-N (Faux et al., 2005).

Homorepeats are preferably located at the N- (43%) and/or C-terminal (30%) flanking regions of well-defined structural domains and they are seldom found within (16%) or between (10%) of the domains (Faux et al., 2005). The flanking regions tend to be disordered (Alba et al., 2007). Poly-L runs mainly occur within the first 40 residues corresponding to the signal-peptide sequence (Karlin et al., 2002). The other homorepeats do not have such a strong preference for specific regions within proteins.

F. Protein Size and Tendency to Have Homorepeats

It was shown that the percentage of proteins containing at least one homorepeat rises with an increase in protein size (Singh et al., 2004). It may appear obvious that it is more probable to find a homorepeat as the length of a protein increases. However, this tendency is also observed when the frequency of homorepeats per unit of length is plotted, but it is less pronounced (Fig. 3). Remarkably, at the beginning of the plot, the smallest proteins (<100 residues) have a high propensity for homorepeats, which drastically decreases for proteins with sizes of about 200 residues. As the protein size increases, the homorepeat propensity gradually rises and reaches its maximum at around 900 residues followed by another maximum at 1,700 residues. Interestingly, according to this plot, proteins of sizes less than 100 residues can be distinguished as a special group by their strong tendency to have homorepeats. Amino acid compositions of these homorepeats support the conclusion that they have distinctive characteristics (Fig. 3). They have un unusually high proportion of positively charged runs (20.8% vs. on the average 1.8% for all proteins) and have almost no negatively charged runs (3.2% vs. 12.1%). The proportion of the hydrophobic runs is unusually high (8.7% vs. 2%). The majority of the proteins containing positively charged runs bind to DNA in the nucleus, while ones with hydrophobic runs are located in the

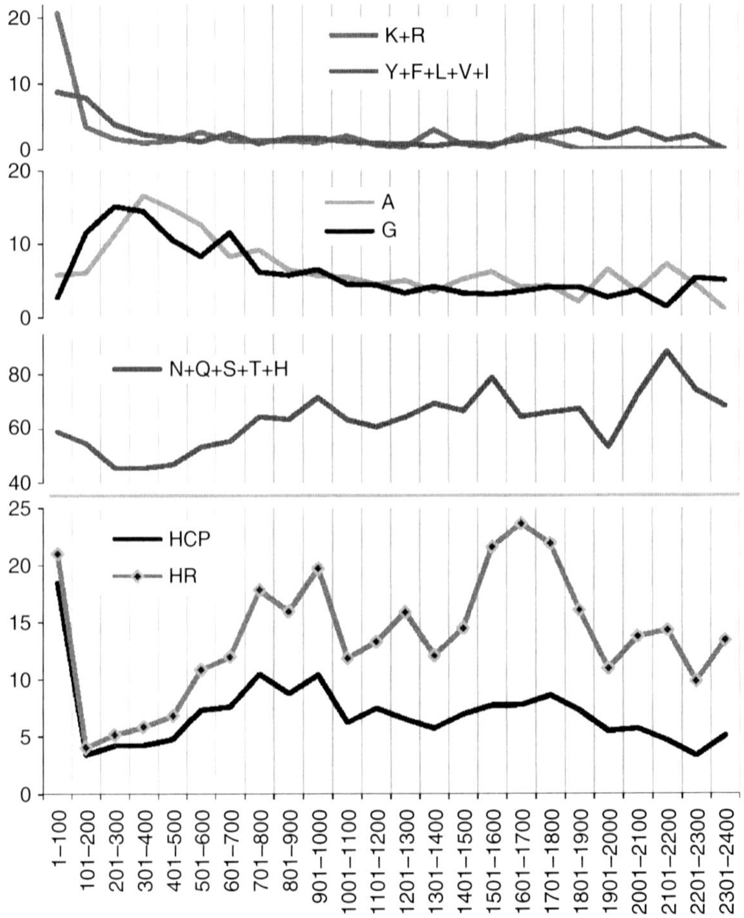

FIG. 3. Frequencies per unit length (in arbitrary units) of Swiss-Prot homorepeat-containing proteins (HCP) and homorepeats (HR) as a function of protein size (at the bottom). Above, corresponding amino acid compositions of HRs are shown.

membranes. Among these small proteins with homorepeats, cytoplasmic proteins are practically absent. Proteins of size between 100 and 600 residues also have biased compositions of homorepeats with increased proportions of poly-A and poly-G and with decreased proportions of polar homorepeats (Fig. 3).

III. Structural Properties of Homorepeat Regions

A. *Synthetic Homopeptides Form Crystallites*

Synthetic homopeptides were among the first objects of structural studies of proteins due to the availability of such specimens and the relative simplicity of data interpretation. It was shown that under physiological conditions, homopeptides containing charged or polar amino acid residues tend to be in a random coil conformation while many nonpolar homopeptides form aggregates (Bloom et al., 1962; Schulz and Schirmer, 1979). Some correlations were found between the types of amino acid residues and the preference of the homopeptides to be in α-helix or β-structure (Bloom et al., 1962; Auer and Doty, 1966). For example, it was noticed that C_β-branched amino acid residues prefer the β-structure to α-helical conformation. Nonpolar solutions that model the environment within biological membranes favor the α-helical conformation for most homopeptides. Studies of homopeptides also revealed their ability to form well-ordered crystallites suitable for X-ray fiber diffraction studies (Fraser and MacRae, 1973). This aptitude of the homopeptides was used to determine the geometry, basic conformation, and interchain packing of the polypeptide chain (Fraser and MacRae, 1973). An important outcome of these studies was a description of the principal elements of the secondary structure found in proteins: the α-helix, the parallel and antiparallel pleated β-sheet structures and the polyproline II helices (Pauling and Corey, 1951; Fraser and MacRae, 1973; Fraser et al., 1973; Eisenberg, 2003) (Fig. 4A and 4B).

Depending on the amino acid type, peptide length, and conditions of the experiment (procedure of crystallization, nature of solution, peptide concentration, ionic strength, pH, temperature), homopeptides can form different types of crystallites. For example, poly-A and poly-K peptides can self-assemble into α-helical or β-structure crystallites (Blout, 1962; Auer and Doty, 1966). Poly-V and poly-S exclusively form the β-structured assemblies while poly-Y peptides favor α-helical structures (Schulz and Schirmer, 1979).

In α-helices, the peptide groups of the backbone are involved in intrahelical H-bonds. The side chains are on the surface and play a critical role in the intermolecular interactions (Fig. 4). In the crystalline form, the α-helices are tightly packed in the hexagonal arrays interacting via the side chains. This favors a platelet-like morphology with hexagonal habit

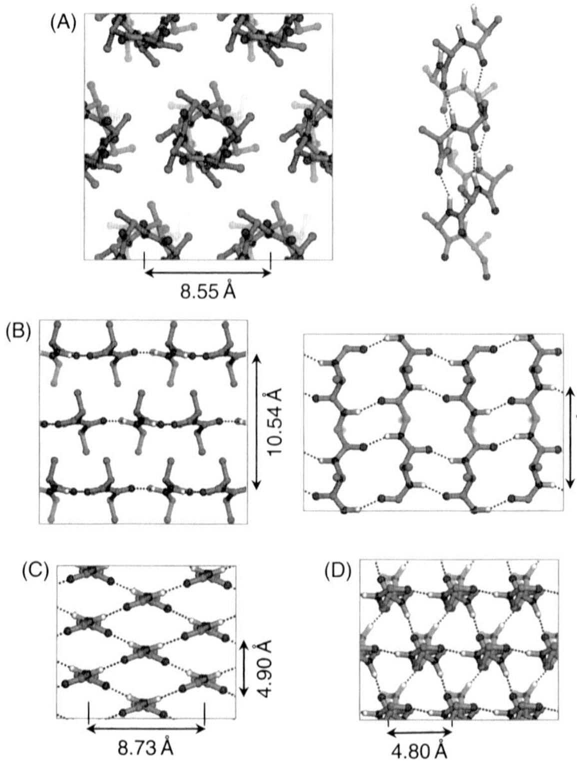

FIG. 4. Principal types of crystal structures formed by homopeptides. The structures are in ball-and-stick representations with carbon atoms depicted in green, nitrogens in blue, and oxygens in red. Only peptide group hydrogens are shown. H-bonds are shown as red dotted lines. (See color plate 4) (A) On the left, axial projection of the crystal structure formed by poly-A α-helices. The hexagonal unit cell has the parameters: $a = 8.55$ Å and $c = 70.3$ Å (Arnott and Dover, 1967). On the right, a lateral view of poly-A α-helix is shown. (B) Axial projection of the antiparallel β-structure of poly-A with an orthogonal unit cell $a = 9.46$ Å, $b = 10.54$ Å, and $c = 6.89$ Å (chain axis) (Arnott et al., 1967). On the right, a lateral view on one antiparallel β-sheet of this crystal structure. (C and D) Axial projections of crystal structures proposed for poly-G form I (Kajava, 1999) and form II (Crick and Rich, 1955). The structure of form I shown in this figure is formed by low molecular weight poly-G molecules. It has a monoclinic unit cell with $a = 4.9$ Å, $b = 8.73$ Å, $c = 7.18$ Å (chain axis), and $\beta = 130.1°$. Form I for high molecular weight molecules is formed by stacking of rippled layers that are similar to β-structural layers. Form II has a hexagonal unit cell ($a = 4.8$ Å, $c = 9.3$, chain axis). The molecular structures were generated with Pymol (De Lano, 2002) (See color plate 5).

(Padden and Keith, 1965). In the α-helical crystallites, all alanines cannot have equivalent environments due to the 3.6 residue per turn helical symmetry. In contrast, within the β-structure crystallites, each residue of a homopeptide has an equivalent environment (Fig. 4). β-Structures are formed by polypeptide chains in extended conformations. The chains interact with each other via peptide group H-bonds and form a flat β-sheet. Within the β-sheets the chains can be arranged in either a parallel or an antiparallel manner. The side chains project above and below the plane of the sheet. The β-sheets are stacked and tightly packed to form stable crystallites. In general, the high stability of these structures is provided by close packing of nonpolar side chains. However, it was also shown that polar poly-Q homopeptides can self-assemble into very stable and high-density β-crystallites (Sikorski and Atkins, 2005) due to the ability of these side chains to form networks of H-bonds between β-sheets (Fig. 5). Given the fact that the expansion of poly-Q runs causes a

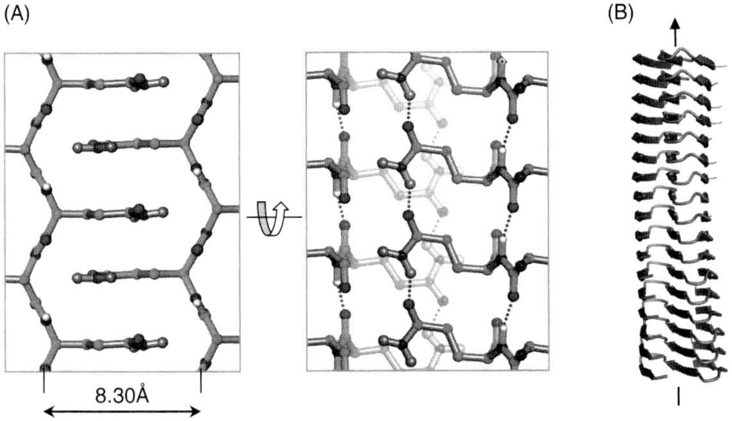

Fig. 5. (A) A model of β-crystallite of poly-Q with an orthogonal unit cell with $a = 9.5$ Å, $b = 16.60$ Å, and $c = 6.95$ Å (chain axis) (Sikorski and Atkins, 2005). On the left, a view along the H-bond directions shows interdigitation of Q side chains. On the right, a view along the β-strands illustrates how the H-bonding of Q side chains creates ladders inside the β-structure. Sticks of the main chains are in green and of the side chains are in gray. (B) A schematic diagram of a cross-β amyloid fibril formed by twisted β-sheets. Beta-strands are shown as arrows. The twist can delimit lateral dimensions and favor axial growth of the homopeptide aggregates. When considered locally, side chain packings (e.g., the H-bonded Q-ladders) within the crystallite (A) and fibril-like assemblies (B) can be similar. (See color plate 5).

number of neurodegenerative disorders, including Huntington's disease (Orr and Zoghbi, 2007), this crystallite structure is important in understanding the role of poly-Q segments in neurodegenerative disorders.

It is worth mentioning that, in known protein structures, β-sheets are rarely flat. Rather, they exhibit a slight twist due to the chirality of their amino acids (Chothia, 1973). This twist delimits the length of the β-strands and the number of β-sheets in the stacks, and by doing so it can favor formation of long fibrous cross-β structures (Nelson and Eisenberg, 2006) (Fig. 5) instead of crystallites.

The crystallites formed by poly-P and poly-G homopeptides differ from both the α-helical and the β-structure crystals. The characteristic properties of poly-P are derived from the fact that Pro residue has a cyclic side chain and its ϕ dihedral angle is fixed at approximately $-75°$ providing exceptional rigidity. On the other hand, it lacks a hydrogen on the amide group and, therefore, cannot form interpeptide H-bonds between poly-P chains. These constraints make poly-P incompatible with the α-helical and β-structure conformations. Two forms of poly-P have been found (Harrington and Sela, 1958): form I consists of regular poly-P helices with the *cis*-configuration of the peptide bonds and form II is a regular helix with the *trans*-peptide groups.

Poly-G also stands apart from other homopeptides. Glycine has a hydrogen atom as its side chain and does not contain an asymmetric C_α-atom. Therefore, the mirror-symmetrical aggregates of poly-G can be formed with equal probability. Furthermore, the small size of glycine allows these polypeptides to approach each other so closely in the crystals that their peptide groups form a spatial network of H-bonds. Poly-G can form at least three crystalline forms. The crystal modification "polyglycine I" has two different forms depending on the molecular weight. In the low molecular weight (~10–15 residues) form I, the molecules have an unusual extended conformation generated by the alternation of two mirror-symmetrical residual conformations along the chain (Kajava, 1999). The molecules are parallel with respect to the direction from N- to C-terminus and each chain forms interpeptide H-bonds with four adjacent chains (Fig. 4). The high molecular weight (~90 residues) form I crystal represents a composition of twinning crystallites. The crystallites themselves consist of antiparallel enantiomorphous chains united by H-bonds to form rippled sheets (Pauling and Corey, 1953; Kajava, 1999). In the high molecular weight form I the orientation of the chains is antiparallel, but it is parallel in the low

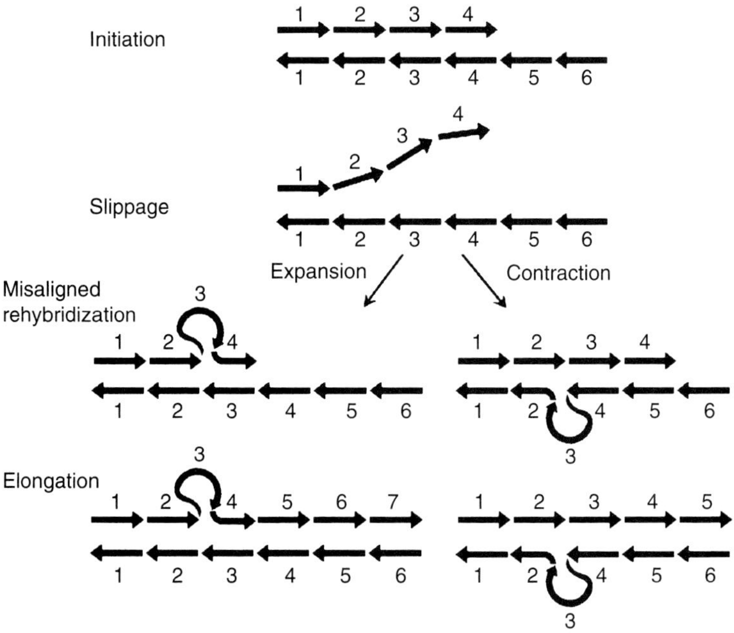

FIG. 6. Model of molecular mechanism of microsatellite mutation based on replication slippage. DNA repeat units are shown by arrows and numbered within each strand.

molecular weight structure. Therefore, one crystal modification cannot be transformed into the other by mechanical or other means. In the crystal form "polyglycine II," each molecule has a threefold screw axis and conformation similar to the poly-P II helix (Crick and Rich, 1955) (Fig. 4). The chains are parallel with respect to the N- to C-terminus direction, packed in a hexagonal array and each is H-bonded to its six nearest neighbors.

B. Living Organisms Avoid Structure-Forming Homorepeats

Self-assembly itself is not toxic for the cell and is widely used by natural proteins. It is not the assembly itself but the inherent property of homorepeat structures to form stable aggregates of unlimited size that may be a basis of homorepeat cytotoxicity. Both properties (unlimited size and high stability) of the homopeptide structures do not correspond to requirements of biological molecules: a) to be able to control their dimensions and b) when it is necessary, to be readily degraded (Kajava, 2001). Thus, the homopeptide structures are predisposed to form deposits that are harmful to living organisms (Perutz, 1999).

When homorepeats with the potential to self-assemble occur in natural proteins, they are usually flanked by large aperiodic sequences typical of globular structures (Kreil and Kreil, 2000; Kajava, 2001). The flanking globular structures may prevent the formation of homorepeat associates by steric impedance or interactions with the homorepeat region. As a rule, those rare proteins that contain homorepeat runs of bulky nonpolar residues target membranes after their biosynthesis and are not released to the aqueous solution. Poly-L, the most frequent among them, is preferentially located at the N-terminal leader peptide region, which is proteolytically digested after translocation of its protein across the membrane (Karlin et al., 2002). Most of the rare proteins that are composed of multiple stretches of structure-forming homorepeats are extracellular. For example, production of silk fibers is designed so that poly-A-containing silk fibroin forms its final structure only when outside the organism (Hinman et al., 2000).

IV. Evolutionary Dynamics of Homorepeats

Homorepeat regions of proteins are coded by tandemly repeated stretches of DNA known as microsatellites (Charlesworth et al., 1994). The microsatellites differ from most other types of DNA sequences in their unusually high polymorphism (Goldstein and Schlötterer, 1999). Depending on the organism, the mutation rate of microsatellites is several orders of magnitude higher (10^{3}–10^{6} per locus per generation) than the rate of the unique DNA sequence (Weber and Wong, 1993; Ellegren, 2000). The rate also depends on the total length of the individual microsatellite—long loci mutate more often than short ones (Wierdl et al., 1997; Brinkmann et al., 1998). A series of studies suggests that the homorepeat instability can be both intergenerational and somatic (Brouwer et al., 2009).

The mutation mechanisms contributing to the length polymorphism at microsatellite loci are still a matter of debate. However, an increasing amount of evidence suggests a model based on replication slippage (Levinson and Gutman, 1987; Ellegren, 2000). According to this model, when slipping-strand mispairing occurs within a microsatellite region during DNA synthesis, it can result in the gain or loss of one or more repeat units depending on whether the newly synthesized DNA chain or the template chain respectively loops out (Fig. 6). Some data suggest that the loops of highly expandable repeats can form hairpin structures with both Watson–Crick and mismatched base pairs (Mirkin, 2006).

The homorepeat regions of proteins consist predominantly of homocodons (the same type of codons) and they do not have an apparent G+C or A+T nucleotide base bias relative to the organism's transcriptome (Faux et al., 2007). In general, transition substitutions within the codons are more common than transversions (Collins and Jukes, 1994; Morton, 1995); however, most of the homorepeats comprise a homocodon interrupted by a single base transversion (Faux et al., 2007). Disruptions of homocodons are likely to reduce the rate of slippage mutation as shown in *S. cerevisiae* (Rolfsmeier and Lahue, 2000), probably by reduction of DNA loop formation during tandem repeat replication.

Thus, on one hand, the homorepeat regions evolve more rapidly than the rest of the gene but, on the other hand (see Section III.C), their expansion may be toxic to the cells. This suggests that the homorepeat expansion and contraction is under a strong control by a number of factors including DNA replication, repair, and modification machineries (Karlin et al., 2002; Cleary and Pearson, 2003; Feschenko et al., 2003). Given the fact that the homorepeat regions are harmful to the organisms, it is unclear as to why the instability of homorepeat regions be maintained throughout evolution? Recently, it was shown that homorepeat instability can have positive phenotypic effects. For example, homorepeat expansion and contraction within *cis*-regulatory elements of developmental genes is linked to morphological changes in dog breeds (Fondon and Garner, 2004). This observation offers the hypothesis that the repeat length instability is a major source of phenotypic variation in evolution (Fondon and Garner, 2004).

V. Functional Associations of Homorepeat Regions

A. *Functions of Proteins Containing Homorepeat Regions*

Several studies have indicated an association between homorepeat-containing proteins and their functional properties (Singh et al., 2004). These proteins are involved in a large variety of functions including catalytic activity, enzyme regulation, transport, channeling, adhesion, signaling, translation, and structural functions. A survey of proteins with homorepeat runs indicates that despite a diversity of functions, they tend to belong to specific functional classes of intracellular proteins, particularly transcription factors and protein kinases (Karlin and Burge, 1996; Alba and Guigo, 2004). This is especially true for eukaryotes. Prokaryotes, on the other hand, do not have such overrepresentation of homorepeat-containing proteins of transcription/translation and signaling. They account for 24% of enzymes and 12% of adhesion molecules among the proteins considered (Faux et al., 2005). In eukaryotes, the majority of the homorepeat-containing proteins are associated with large multiprotein and/or nucleic acid complexes such as ribosomes and their factors, complexes involved in transcriptions and chromatin condensation and signaling complexes (e.g., MAP kinase) (Faux et al., 2005). According to the PRDB, in viruses, the greatest number of homorepeats is found among structural proteins of capsids and envelopes followed by transcriptional factors.

B. Do Homorepeats Represent Junk Protein Regions or Have Essential Functional Roles?

Attempts to find a specific functional role for protein homorepeats have produced contradictory conclusions. Some authors suggest that the homorepeats serve primarily as spacers between functionally important domains (Huntley and Golding, 2000). This point of view is based on the facts that a) the majority of the homorepeat runs are predicted to be intrinsically unfolded (or unstructured) (Huntley and Golding, 2002), b) the rate of evolution is higher within than outside of the homorepeat, c) the repeat region length as well as type of amino acid residue can vary between species (Newfeld et al., 1994) and, d) homorepeats of some proteins may be deleted without affecting the function (Gatti et al., 1994; Sumiyama et al., 1996). At the same time, other studies clearly demonstrate that homorepeats themselves are critical to the function of some repeat-containing proteins. For example, stretches of poly-Q or poly-P can activate transcription when fused to the DNA-binding domain of GAL4 factor (Gerber et al., 1994). The transcriptional activation can be modulated by variation of the length and amino acid type of the homorepeats. Comparative genomics and morphological studies suggested that length variation in homorepeats can be a major source of morphological variations in dog breeds (Fondon and Garner, 2004). The role of neither poly-A nor poly-G runs that are found in extracellular proteins such as silk fibroin and elastin can be reduced to spacers between the domains (Altman et al., 2003).

The most abundant homorepeats of poly-Q and poly-N are prone to cytotoxic aggregation. However, they were not replaced during evolution by less harmful hydrophilic homorepeats. This suggests that poly-Q or poly-N may be involved in specific binding with other molecules rather than serving as a neutral cross-link between functional domains.

These and other examples clearly demonstrate the diversity of homorepeat functional roles rather than the existence of one function. The frequent absence of a stable folded structure of homorepeat regions (Huntley and Golding, 2002) cannot be used as a proof in favor of the insignificance of the homorepeat functions. The intrinsically unfolded regions have a number of important functions (Radivojac et al., 2007). At the same time, often, expansion or contraction of the homorepeat regions provide subtle regulation and modulation of the basic function of their proteins (Gerber et al., 1994; Fondon and Garner, 2004). Therefore, the rapid evolution of these regions (Newfeld et al., 1994) should not endanger the population but rather provide means of attaining the necessary diversity in the face of evolutionary pressure.

C. Preferences of Homorepeat Types for Certain Localizations and Functions

Homorepeats of some amino acid residues occur more frequently than others depending on the function or subcellular localization of their proteins. According to the PRDB, almost all proteins containing poly-R are located in the nucleus. This relationship also holds, but is less pronounced, in proteins containing poly-K. This observation is supported by experiments with recombinant proteins fused with poly-R or poly-K tracts that are found in the nucleus (Nam et al., 2001; Oma et al., 2004). Poly-K runs often occur in membrane proteins. In contrast, the negatively charged poly-E and poly-D are abundant among cytoplasmic proteins (Fig. 7). The cytoplasmic proteins are also enriched in poly-P and poly-G. Nuclear proteins have a high proportion of poly-A and a low proportion of polar homorepeats. As expected, membrane proteins are overrepresented by nonpolar runs. Nonpolar runs are also frequent in secreted proteins.

Clustering of homorepeat-containing proteins by their functions also reveals preferences for certain types of amino acid residues (Fig. 7). For example, the occurrence of poly-A runs is greater-than-average in transcription factors while negatively and (more surprisingly) positively charged residues are almost absent among these proteins. Structural proteins have many poly-G and poly-A, but very few polar homorepeat runs.

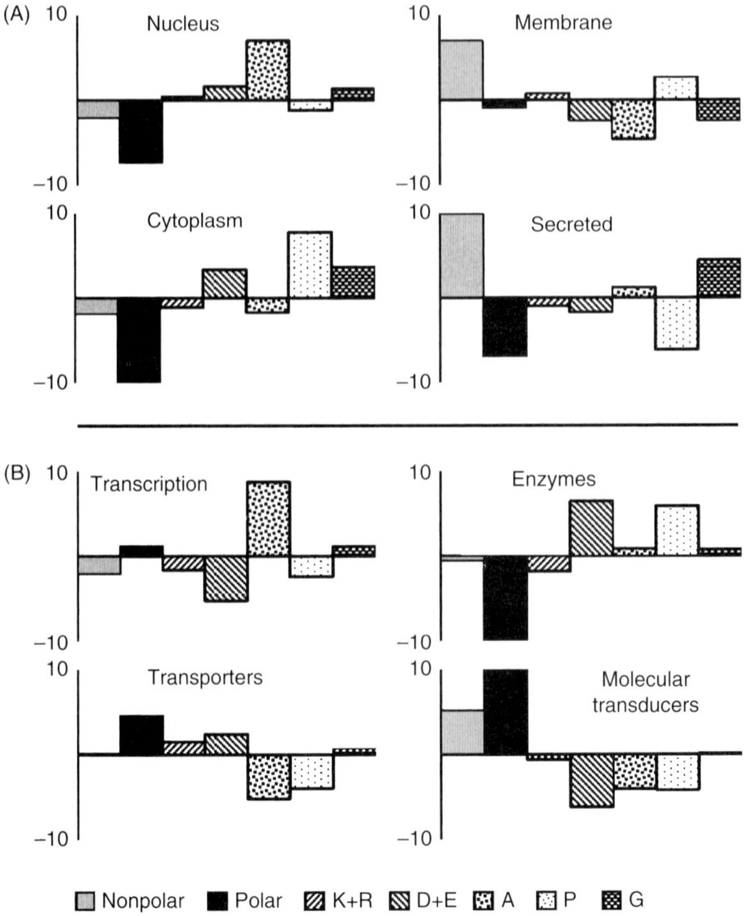

FIG. 7. Histograms presenting differences between frequencies of types of homorepeats assigned to particular localizations (A) and functions (B) and the average frequency of all homorepeats from Swiss-Prot.

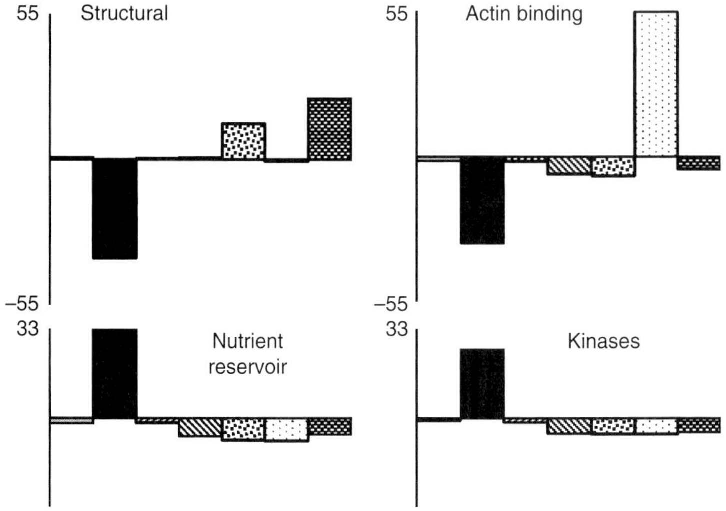

FIG. 7. (Continued)

In contrast, kinases and proteins involved in nutrient reservoir activity have an extremely high proportion of polar homorepeats. Actin-binding proteins have an unusually high content of poly-P runs and a low proportion of polar homorepeats. This observation correlates with the ability of several actin-binding proteins (profilin, SH3, WW-domains, and others) to bind proline-rich regions. Enzymes have many negatively charged runs and poly-P, but a "less-than-average" proportion of polar runs.

D. *Functions Specifically Related to Homorepeat Regions*

Usually, homorepeats occur in large proteins containing several functional domains. Therefore, it is difficult to separate the homorepeat function from the remaining part of the protein. Knowledge of the unique properties of homorepeat sequences, structures, and evolution (see previous sections) in combination with the functional data and mutational analysis of the repeat-containing proteins allows us to speculate about functions that are specifically linked to homorepeats.

The exact repetitiveness and homogeneity of the sequences suggest that the homorepeats can interact well in a multivalent manner with partners having regular structures, such as DNA, RNA, carbohydrates, repetitive protein structures, or nonorganic crystals (Fig. 8). In several cases when a homorepeat-containing protein interacts with DNA, the variation of the homorepeat length provides subtle modulation of the basic protein function (Gerber et al., 1994; Fondon and Garner, 2004). This suggests that the interactions of the homorepeat with DNA are nonspecific and are characterized by weak energy contributions per amino acid residue. DNA grooves may provide such regular binding surfaces for the homorepeat regions. Interactions are likely between the positively charged homorepeats and the negatively charged backbone of the nucleic acid. It can also be mediated by H-bonding between the DNA/RNA grooves and poly-Q, poly-N, poly-A, poly-P, or poly-G in extended conformations. The fact that different types of homorepeats differ in their optimal length for transcriptional activation (Gerber et al., 1994) can be explained by the amino acid residue-specific differences in the strengths of interaction with DNA. The mode of a linear, nonspecific interaction can tolerate a certain level of homorepeat length variation

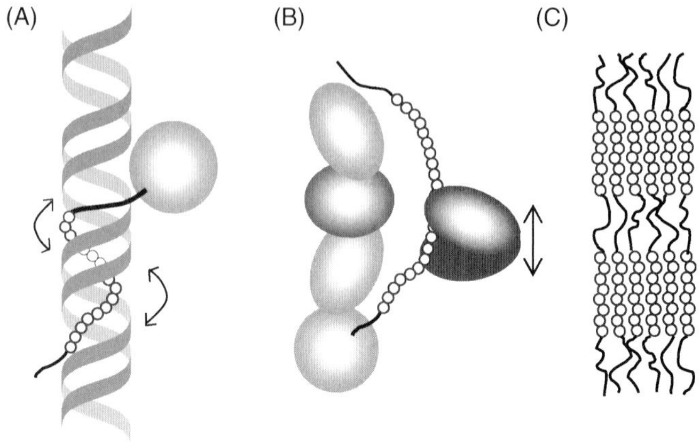

FIG. 8. Examples of homorepeat-specific functions. Globular protein structures are shown as spheres and homorepeat regions as "beads on string." (A) DNA–homopeptide interaction is shown as an example of multivalent binding to repetitive structures. (B) Sliding of molecules along homorepeat regions. This movement can provide necessary mobility to subunits of a supramolecular machine. (C) Formation of homorepeat crystalline structures.

quite well. Therefore, it can also efficiently attenuate the evolutionary instability of these regions.

The exact repetitiveness of the homorepeats also provides a basis for efficient interaction with nonorganic crystals or catalysis of their crystallization. For example, poly-E runs of bone sialoprotein are involved in the hydroxyapatite-nucleating activity (Tye et al., 2003).

The homogeneity of the homorepeats can also provide another set of unique functional properties. It is known that homorepeat-containing proteins participate in large suparmolecular machines and mediate and regulate protein–protein interaction networks (Remacha et al., 1995; Korschen et al., 1999; Mayer, 2001; Poetsch et al., 2001; Hancock and Simon, 2005). When the size of a homorepeat stretch is larger than its binding site to a protein partner, the bound protein can slide along the homorepeat region while maintaining contact. *Sliding* along the homorepeat region may impart dynamic properties to the supramolecular machine, which can be critical for its functioning (Fig. 8). Such homorepeats may tolerate the extension–contraction by a few residues that may occur during evolution. Complexes between homorepeats and nucleic acid molecules can also be very dynamic due to the sliding mechanism. The sliding might be critical for proper functioning of large protein–nucleic acid complexes involved in transcription, translation, and recombination processes.

The homogeneity of homorepeats provides a predisposition to interact not only with other periodic molecules (e.g. DNA, RNA) but also with themselves forming crystalline structures or amyloid fibrils (Fig. 8). Frequently, this aggregation is associated with cytotoxicity. However, nature also uses this property for beneficial biological functions. For example, the homorepeat crystallites are required to stiffen silk fibers or elastin tissue (Hinman et al., 2000; Altman et al., 2003). Furthermore, recent studies have identified amyloid fibers in bacteria, fungi, insects, invertebrates, and humans that are necessary for the proper functioning (Fowler et al., 2007; Maji et al., 2009).

Previously it was proposed that homorepeats, due to their inherent predispositions to rapid evolution, may be used to increase the overall size of proteins (Lavorgna et al., 2001). When such a possibility exists, we believe that it represents a rather minor, collateral role for homorepeats. It does not require specific properties of the homorepeats and can be fulfilled even more efficiently by multiplication of longer (more than

1 residue) repetitive units. The principal role of the homorepeats as a flexible neutral linker between folded domains (Karlin and Burge, 1996; Faux et al., 2005) is also unlikely. This function does not require perfect repetitiveness of the link and can be provided by other less regular and intrinsically unfolded regions. The fact that only a minor portion (about 10%) of homorepeat regions occurs between folded functional domains (Faux et al., 2005) supports this conclusion.

E. Function of a Homorepeat Region Strongly Depends on the Adjacent Domains and Subcellular Localization of the Protein

In contrast to the globular domains, homorepeats of the same type and length can have very different functions depending on the adjacent domains and subcellular localization of the protein. For example, runs of poly-A or poly-G fused with DNA-binding domains of transcriptional factors can modulate activity of these proteins (Gerber et al., 1994). On the other hand, the same poly-A or poly-G as a part of silk fibroin or elastin tend to self-assemble into the crystallites that are necessary to support the mechanical properties of the proteins (Fraser and MacRae, 1973; Altman et al., 2003). Depending on the protein context, poly-P homorepeat can serve as cross-links, modulate transcriptional activity (Gerber et al., 1994), or provide a binding site for proteins (e.g., SH3 domain) interactions (Mayer, 2001). Thus, susceptibility to flanking domains and cellular localization are important factors that should be taken into consideration in predicting homorepeat functions.

F. Homopeptides as Building Blocks of New Materials and Supramolecular Nanostructures

Some extracellular proteins such as silk fibroin, elastin, plant cell wall proteins, and mollusc shell framework proteins contain runs of structure-forming homorepeats such as poly-A, poly-G, or poly-P. Structural studies also revealed that two crystal modifications of polyglycine molecules are relevant to artificial nylon materials (Tormo et al., 1992; Kajava, 1999). Synthetic polyglycine-containing molecules can self-assemble in an aqueous medium into supramolecular structures of different shapes and properties ranging from plane layers to vesicles (Bovin et al., 2008). This opens new avenues in the development of technology designed to coat surfaces with biocompatible materials, and the creation of supramolecular structures as antiviral agents or drug-delivery vectors. Thus, because of their impressive self-assembly properties and mechanical characteristics, homorepeat runs can provide an important set of material options in the fields of controlled release, biomaterials, and scaffolds for tissue engineering (Altman et al., 2003).

Acknowledgments

The authors thank A. Ahmed, A. Nesterova, and T. Vasselon for critical reading of the manuscript and suggestions. This work was supported by a doctoral fellowship to J.J. from Ministère de l'Education Nationale, de la Recherche et de la Technologie.

References

Ai, H., Lvov, Y. M., Mills, D. K., Jennings, M., Alexander, J. S., Jones, S. A., (2003). Coating and selective deposition of nanofilm on silicone rubber for cell adhesion and growth. *Cell Biochem. Biophys.* **38**, 103–114.

Alba, M. M., Guigo, R., (2004). Comparative analysis of amino acid repeats in rodents and humans. *Genome Res.* **14**, 549–554.

Alba, M. M., Tompa, P., Veitia, R. A., (2007). Amino acid repeats and the structure and evolution of proteins. *Genome Dyn.* **3**, 119–130.

Altman, G. H., Diaz, F., Jakuba, C., Calabro, T., Horan, R. L., Chen, J., et al., (2003). Silk-based biomaterials. *Biomaterials* **24**, 401–416.

Arnott, S., Dover, S.D., Elliott, A., (1967). Structure of beta-poly-L-alanine: refined atomic co-ordinates for an anti-parallel beta-pleated sheet. *J. Mol. Biol* **30**, 201–208.

Arnott, S., Dover, S.D., (1967). Refinement of bond angles of an alpha-helix. *J. Mol. Biol* **30**, 209–212.

Auer, H. E., Doty, P., (1966). The conformational stability of alpha-helical nonpolar polypeptides in solution. *Biochemistry* **5**, 1716–1725.

Bloom, S. M., Fasman, G. D., de Lozé, C., Blout, E. R., (1962). The effect of amino acid composition on the conformations of synthetic polypeptides, polymers and copolymers of L-methionine S-methyl-L-cysteine and L-valine. *J. Am. Chem. Soc.* **84**, 458–463.

Blout, E. R., (1962). Polyamino acids, polypeptides and proteins. In: Stahmann, M. A. (Eds.), University of Wisconsin Press, Madison, p. 275.

Bovin, N. V., Tuzikov, A. B., Chinarev, A. A., (2008). Oligoglycines: materials with unlimited potential for nanotechnologies. *Nanotechnol. Russ.* **3**, 291–302.

Brais, B., Bouchard, J. P., Xie, Y. G., Rochefort, D. L., Chretien, N., Tome, F. M., et al., (1998). Short GCG expansions in the PABP2 gene cause oculopharyngeal muscular dystrophy. *Nat. Genet.* **18**, 164–167.

Brinkmann, B., Klintschar, M., Neuhuber, F., Huhne, J., Rolf, B., (1998). Mutation rate in human microsatellites: influence of the structure and length of the tandem repeat. *Am. J. Hum. Genet.* **62**, 1408–1415.

Brouwer, J. R., Willemsen, R., Oostra, B. A., (2009). Microsatellite repeat instability and neurological disease. *Bioessays* **31**, 71–83.

Charlesworth, B., Sniegowski, P., Stephan, W., (1994). The evolutionary dynamics of repetitive DNA in eukaryotes. *Nature* **371**, 215–220.

Chothia, C., (1973). Conformation of twisted beta-pleated sheets in proteins. *J. Mol. Biol.* **75**, 295–302.

Cleary, J. D., Pearson, C. E., (2003). The contribution of cis-elements to disease-associated repeat instability: clinical and experimental evidence. *Cytogenet. Genome Res.* **100**, 25–55.

Collins, D. W., Jukes, T. H., (1994). Rates of transition and transversion in coding sequences since the human-rodent divergence. *Genomics* **20**, 386–396.

Crick, F. H., Rich, A., (1955). Structure of polyglycine II. *Nature* **176**, 780–781.

Csuros, M., Noe, L., Kucherov, G., (2007). Reconsidering the significance of genomic word frequencies. *Trends Genet.* **23**, 543–546.

DeLano, W.,L., (2002). The PyMOL Molecular Graphics System, DeLano Scientific, San Carlos, CA.

Edde, B., Rossier, J., Le Caer, J. P., Desbruyeres, E., Gros, F., Denoulet, P., (1990). Posttranslational glutamylation of alpha-tubulin. *Science* **247**, 83–85.

Eisenberg, D., Schwarz, E., Komaromy, M., Wall, R., (1984). Analysis of membrane and surface protein sequences with hydrophobic moment plot. *J. Mol. Biol.* **179**, 125–142.

Eisenberg, D., (2003). The discovery of the alpha-helix and beta-sheet, the principal structural features of proteins. *Proc. Natl. Acad. Sci. U.S.A.* **100**, 11207–11210.

Ellegren, H., (2000). Microsatellite mutations in the germline: implications for evolutionary inference. *Trends Genet.* **16**, 551–558.

Fan, X., Dion, P., Laganiere, J., Brais, B., Rouleau, G. A., (2001). Oligomerization of polyalanine expanded PABPN1 facilitates nuclear protein aggregation that is associated with cell death. *Hum. Mol. Genet.* **10**, 2341–2351.

Faux, N. G., Bottomley, S. P., Lesk, A. M., Irving, J. A., Morrison, J. R., de la Banda, M. G., et al., (2005). Functional insights from the distribution and role of homopeptide repeat-containing proteins. *Genome Res.* **15**, 537–551.

Faux, N. G., Huttley, G. A., Mahmood, K., Webb, G. I., de la Banda, M. G., Whisstock, J. C., (2007). RCPdb: an evolutionary classification and codon usage database for repeat-containing proteins. *Genome Res.* **17**, 1118–1127.

Feschenko, V. V., Rajman, L. A., Lovett, S. T., (2003). Stabilization of perfect and imperfect tandem repeats by single-strand DNA exonucleases. *Proc. Natl. Acad. Sci. U.S.A.* **100**, 1134–1139.

Fondon, J. W. 3rd, Garner, H. R., (2004). Molecular origins of rapid and continuous morphological evolution. *Proc. Natl. Acad. Sci. U.S.A.* **101**, 18058–18063.

Fowler, D. M., Koulov, A. V., Balch, W. E., Kelly, J. W., (2007). Functional amyloid–from bacteria to humans. *Trends Biochem. Sci.* **32**, 217–224.

Fraser, R. D., Gillespie, J. M., Macrae, T. P., (1973). Tyrosine-rich proteins in keratins. *Comp. Biochem. Physiol. B* **44**, 943–947.

Fraser, R. D.B., MacRae, T. P., (1973). Conformation in Fibrous Proteins and Related Synthetic Polypeptides. Academic Press, London and New York.

Gatti, E., Popolo, L., Vai, M., Rota, N., Alberghina, L., (1994). O-linked oligosaccharides in yeast glycosyl phosphatidylinositol-anchored protein gp115 are clustered in a serine-rich region not essential for its function. *J. Biol. Chem.* **269**, 19695–19700.

Gerber, H. P., Seipel, K., Georgiev, O., Hofferer, M., Hug, M., Rusconi, S., et al., (1994). Transcriptional activation modulated by homopolymeric glutamine and proline stretches. *Science* **263**, 808–811.

Goldstein, D. B., Schlötterer, C., (1999). Microsatellites Evolution and Applications. Oxford University Press, New York.

Green, H., Wang, N., (1994). Codon reiteration and the evolution of proteins. *Proc. Natl. Acad. Sci. U.S.A.* **91**, 4298–4302.

Hamosh, A., Scott, A. F., Amberger, J., Bocchini, C., Valle, D., McKusick, V. A., (2002). Online Mendelian inheritance in man (OMIM), a knowledgebase of human genes and genetic disorders. *Nucleic Acids Res.* **30**, 52–55.

Hancock, J. M., Simon, M., (2005). Simple sequence repeats in proteins and their significance for network evolution. *Gene* **345**, 113–118.

Harrington, W. F., Sela, M., (1958). Studies on the structure of poly-L-proline in solution. *Biochim. Biophys. Acta* **27**, 24–41.

Hengen, P., (1995). Purification of his-tag fusion proteins from *Escherichia coli*. *Trends Biochem. Sci.* **20**, 285–286.

Hinman, M. B., Jones, J. A., Lewis, R. V., (2000). Synthetic spider silk: a modular fiber. *Trends Biotechnol.* **18**, 374–379.

Huntley, M., Golding, G. B., (2000). Evolution of simple sequence in proteins. *J. Mol. Evol.* **51**, 131–140.

Huntley, M. A., Golding, G. B., (2002). Simple sequences are rare in the protein data bank. *Proteins* **48**, 134–140.

Janke, C., Rogowski, K., Wloga, D., Regnard, C., Kajava, A. V., Strub, J. M., et al., (2005). Tubulin polyglutamylase enzymes are members of the TTL domain protein family. *Science* **308**, 1758–1762.

Jorda, J., Kajava, A. V., (2009). T-REKS: identification of tandem REpeats in sequences with a K-meanS based algorithm. *Bioinformatics* **25**, 2632–2638.

Kajava, A. V., (1999). Dimorphism of polyglycine I: structural models for crystal modifications. *Acta Crystallogr. D Biol. Crystallogr.* **55**, 436–442.

Kajava, A. V., (2001). Review: proteins with repeated sequence–structural prediction and modeling. *J. Struct. Biol.* **134**, 132–144.

Karlin, S., (1995). Statistical significance of sequence patterns in proteins. *Curr. Opin. Struct. Biol.* **5**, 360–371.

Karlin, S., Brocchieri, L., Bergman, A., Mrazek, J., Gentles, A. J., (2002). Amino acid runs in eukaryotic proteomes and disease associations. *Proc. Natl. Acad. Sci. U.S.A.* **99**, 333–338.

Karlin, S., Burge, C., (1996). Trinucleotide repeats and long homopeptides in genes and proteins associated with nervous system disease and development. *Proc. Natl. Acad. Sci. U.S.A.* **93**, 1560–1565.

Karlin, S., Chen, C., Gentles, A. J., Cleary, M., (2002). Associations between human disease genes and overlapping gene groups and multiple amino acid runs. *Proc. Natl. Acad. Sci. U.S.A.* **99**, 17008–17013.

Katti, M. V., Ranjekar, P. K., Gupta, V. S., (2001). Differential distribution of simple sequence repeats in eukaryotic genome sequences. *Mol. Biol. Evol.* **18**, 1161–1167.

Korschen, H. G., Beyermann, M., Muller, F., Heck, M., Vantler, M., Koch, K. W., et al., (1999). Interaction of glutamic-acid-rich proteins with the cGMP signalling pathway in rod photoreceptors. *Nature* **400**, 761–766.

Kreil, D. P., Kreil, G., (2000). Asparagine repeats are rare in mammalian proteins. *Trends Biochem. Sci.* **25**, 270–271.

Laumonnier, F., Ronce, N., Hamel, B. C., Thomas, P., Lespinasse, J., Raynaud, M., et al., (2002). Transcription factor SOX3 is involved in X-linked mental retardation with growth hormone deficiency. *Am. J. Hum. Genet.* **71**, 1450–1455.

Lavorgna, G., Patthy, L., Boncinelli, E., (2001). Were protein internal repeats formed by "bricolage"? *Trends Genet.* **17**, 120–123.

Levinson, G., Gutman, G. A., (1987). Slipped-strand mispairing: a major mechanism for DNA sequence evolution. *Mol. Biol. Evol.* **4**, 203–221.

Maji, S. K., Perrin, M. H., Sawaya, M. R., Jessberger, S., Vadodaria, K., Rissman, R. A., et al., (2009). Functional amyloids as natural storage of peptide hormones in pituitary secretory granules. *Science* **325**, 328–332.

Marcotte, E. M., Pellegrini, M., Yeates, T. O., Eisenberg, D., (1999). A census of protein repeats. *J. Mol. Biol.* **293**, 151–160.

Mayer, B. J., (2001). SH3 domains: complexity in moderation. *J. Cell. Sci.* **114**, 1253–1263.

Mirkin, S. M., (2006). DNA structures, repeat expansions and human hereditary disorders. *Curr. Opin. Struct. Biol.* **16**, 351–358.

Morton, B. R., (1995). Neighboring base composition and transversion/transition bias in a comparison of rice and maize chloroplast noncoding regions. *Proc. Natl. Acad. Sci. U.S.A.* **92**, 9717–9721.

Nam, Y. S., Petrovic, A., Jeong, K. S., Venkatesan, S., (2001). Exchange of the basic domain of human immunodeficiency virus type 1 rev for a polyarginine stretch expands the RNA binding specificity, and a minimal arginine cluster is required for optimal RRE RNA binding affinity, nuclear accumulation, and trans-activation. *J. Virol.* **75**, 2957–2971.

Nelson, R., Eisenberg, D., (2006). Structural models of amyloid-like fibrils. *Adv. Protein Chem.* **73**, 235–282.

Newfeld, S. J., Tachida, H., Yedvobnick, B., (1994). Drive-selection equilibrium: homopolymer evolution in the Drosophila gene mastermind. *J. Mol. Evol.* **38**, 637–641.

Oma, Y., Kino, Y., Sasagawa, N., Ishiura, S., (2004). Intracellular localization of homopolymeric amino acid-containing proteins expressed in mammalian cells. *J. Biol. Chem.* **279**, 21217–21222.

Orr, H. T., Zoghbi, H. Y., (2007). Trinucleotide repeat disorders. *Annu. Rev. Neurosci.* **30**, 575–621.

Padden, F. J., Keith, H. D., (1965). *J. Appl. Phys.* **36**, 2987–2995.

Pauling, L., Corey, R. B., (1951). The structure of fibrous proteins of the collagen-gelatin group. *Proc. Natl. Acad. Sci. U.S.A.* **37**, 272–281.

Pauling, L., Corey, R. B., (1953). Two rippled-sheet configurations of polypeptide chains, and a note about the pleated sheets. *Proc. Natl. Acad. Sci. U.S.A.* **39**, 253–256.

Perutz, M. F., (1999). Glutamine repeats and neurodegenerative diseases: molecular aspects. *Trends Biochem. Sci.* **24**, 58–63.

Poetsch, A., Molday, L. L., Molday, R. S., (2001). The cGMP-gated channel and related glutamic acid-rich proteins interact with peripherin-2 at the rim region of rod photoreceptor disc membranes. *J. Biol. Chem.* **276**, 48009–48016.

Radivojac, P., Iakoucheva, L. M., Oldfield, C. J., Obradovic, Z., Uversky, V. N., Dunker, A. K., (2007). Intrinsic disorder and functional proteomics. *Biophys. J.* **92**, 1439–1456.

Redeker, V., Levilliers, N., Schmitter, J. M., Le Caer, J. P., Rossier, J., Adoutte, A., et al., (1994). Polyglycylation of tubulin: a posttranslational modification in axonemal microtubules. *Science* **266**, 1688–1691.

Remacha, M., Jimenez-Diaz, A., Bermejo, B., Rodriguez-Gabriel, M. A., Guarinos, E., Ballesta, J. P., (1995). Ribosomal acidic phosphoproteins P1 and P2 are not required for cell viability but regulate the pattern of protein expression in *Saccharomyces cerevisiae*. *Mol. Cell. Biol.* **15**, 4754–4762.

Rolfsmeier, M. L., Lahue, R. S., (2000). Stabilizing effects of interruptions on trinucleotide repeat expansions in *Saccharomyces cerevisiae*. *Mol. Cell. Biol.* **20**, 173–180.

Schulz, G. E., Schirmer, R. H., (1979). Principles of Protein Structure. Springer-Verlag, New York.

Sikorski, P., Atkins, E., (2005). New model for crystalline polyglutamine assemblies and their connection with amyloid fibrils. *Biomacromolecules* **6**, 425–432.

Singh, G. P., Chandra, B. R., Bhattacharya, A., Akhouri, R. R., Singh, S. K., Sharma, A., (2004). Hyper-expansion of asparagines correlates with an abundance of proteins with prion-like domains in *Plasmodium falciparum*. *Mol. Biochem. Parasitol.* **137**, 307–319.

Siwach, P., Ganesh, S., (2008). Tandem repeats in human disorders: mechanisms and evolution. *Front. Biosci.* **13**, 4467–4484.

Stromme, P., Mangelsdorf, M. E., Shaw, M. A., Lower, K. M., Lewis, S. M., Bruyere, H., et al., (2002). Mutations in the human ortholog of aristaless cause X-linked mental retardation and epilepsy. *Nat. Genet.* **30**, 441–445.

Sumiyama, K., Washio-Watanabe, K., Saitou, N., Hayakawa, T., Ueda, S., (1996). Class III POU genes: generation of homopolymeric amino acid repeats under GC pressure in mammals. *J. Mol. Evol.* **43**, 170–178.

Tormo, J., Puiggali, J., Vives, J., Fita, I., Lloveras, J., Bella, J., et al., (1992). Crystal structure of a helical oligopeptide model of polyglycine II and of other polyamides: acetyl-(glycyl-beta-alanyl)2-NH propyl. *Biopolymers* **32**, 643–648.

Tye, C. E., Rattray, K. R., Warner, K. J., Gordon, J. A., Sodek, J., Hunter, G. K., et al., (2003). Delineation of the hydroxyapatite-nucleating domains of bone sialoprotein. *J. Biol. Chem.* **278**, 7949–7955.

Weber, J. L., Wong, C., (1993). Mutation of human short tandem repeats. *Hum. Mol. Genet.* **2**, 1123–1128.

Wierdl, M., Dominska, M., Petes, T. D., (1997). Microsatellite instability in yeast: dependence on the length of the microsatellite. *Genetics* **146**, 769–779.

PROTEIN AGGREGATION: FROM INCLUSION BODIES TO AMYLOID AND BIOMATERIALS

By ANNA MITRAKI

Department of Materials Science and Technology, University of Crete and Institute for Electronic Structure and Laser, Foundation for Research and Technology-Hellas (IESL-FORTH), Vassilika Vouton, Heraklion, Crete, Greece

Abstract

Inclusion bodies and amyloid are two different outcomes of the same fundamental biological process: protein aggregation. They share two important features that suggest sequence-specific aggregation: oligomeric intermediates as precursors to the aggregated sate and sensitivity to site-specific point mutations. For a long time, the physical state of inclusion bodies was refractory to the use of protein structural characterization techniques that were developed for soluble proteins. Recent high-resolution studies reveal that the apparently amorphous state of these dense protein agglomerates consists of amyloid-type structures adopted by short segments that initiate aggregation. Under certain circumstances it is possible for the aggregation-prone segments to "recruit" a globular counterpart within the inclusion body, with the latter being able to fold into an active conformation. In this chapter we will discuss these recent structural insights in relation with the also recent, high-resolution structures for amyloid "spines" and their ability to accommodate globular or "swapped" domains in their periphery. Finally, unexpected natural roles

DOI: 10.1016/S1876-1623(10)79003-9

for amyloid structures such as protection, adhesion, and storage materials gradually emerge. We will discuss how these properties in combination with biochemical and structural insights can inspire "biomimetic" approaches for the rational design of novel nanobiomaterials.

I. Historical Perspective: "Unboiling the Egg"

"Protein coagulation, together with combustion and fermentation, was one of the first chemical processes recognized by man" pointed M.L. Anson and A.E. Mirsky in their 1931 article entitled "The reversibility of protein coagulation" (Anson and Mirsky, 1931). Indeed, detection of the proteinaceous nature of a natural tissue component or material was based on its distinct ability to coagulate when heated. In the abovementioned article, Anson and Mirsky, by carrying out very simple experiments, have proven that protein coagulation can be avoided with careful choice of conditions. For instance, if acid-denatured hemoglobin or albumin is neutralized in one step by adding at once the equivalent amount of alkali, then the protein will precipitate. However, if an amount of alkali just insufficient to neutralize is added, and the protein is then left under these conditions for a few hours then by subsequent addition of the remaining alkali only a small fraction will precipitate and the rest will remain in solution. The protein that has been "reversed" recovered the structural characteristics of its native form. In an article entitled "Unboiling the egg" published in "Discovery" in 1940, Max Perutz further developed this insight (Perutz, 1980). If one would only be able to physically "unscramble" the coagulated protein, for example, by using denaturing conditions, then it should be possible to return back to the native state by choosing the proper conditions. In an attempt to examine the structure of the coagulated state, W.T. Astbury and colleagues recorded the X-ray diffraction pattern of stretched, boiled egg white strips (Astbury et al., 1935). The pattern displayed two characteristic reflections: one at 4.65 Å assigned to "backbone spacing" and one at 9.8 Å assigned to "side chain spacing." The unexpected result was that the backbone spacing was lying on the meridian and the side chain spacing on the equator. The pattern suggested that the albumin chains were adopting an extended beta-structure arrangement, with the polypeptide chains located perpendicular to the axis of stretching, subsequently termed "cross-beta." In the last 40 years, it became evident that protein aggregation is associated with inclusion body formation during the recombinant production of proteins.

It is also a hallmark of "amyloid" diseases—diseases associated with protein deposition in insoluble, fibrillar forms with the characteristic "cross-beta" signature. More recently, the notion of protein aggregates/amyloids that may play a beneficial functional or structural role in organisms, termed "natural amyloids," has emerged. Furthermore, amyloid structures can be beneficially exploited for bio-nanomaterials design. Disease-associated amyloids and functional amyloids have been the subject of several extensive and excellent reviews. We will therefore focus in this chapter on two less-well explored areas of aggregation phenomena: inclusion bodies and amyloid structures as protein biomaterials.

II. Protein Aggregation *In Vitro*

In the years that followed the seminal success of Anfinsen in renaturing pancreatic ribonuclease (RNase) (Anfinsen, 1973), denaturation–renaturation studies were carried out *in vitro* for many small proteins. Their unfolding–refolding processes were apparently complying with a two-state model, and as a result any deviations were considered to be due to aberrant species or kinetic traps. With the increasing number of larger proteins being studied, the important role of partially folded intermediates emerged (Kim and Baldwin, 1982). Furthermore, only partial renaturation yields were observed, ranging from small percentages to about 90%; accumulation of the protein chains in insoluble, inactive forms was frequently observed. However, protein aggregation was regarded as side phenomenon that did not deserve any scientific investigation by protein biochemists, with only a few exceptions. Rainer Jaenicke and colleagues studied thoroughly the refolding pathways of several oligomeric enzymes and established the competing character of refolding and aggregation pathways (Jaenicke, 1987). Generally, partially folded monomers are transiently populated and can proceed to form either a monomer competent for correct association and oligomerization, or alternatively, off-pathway aggregates (Zettlmeissl et al., 1979, 1981). The aggregation pathways become favored with increasing protein concentration (Rudolph et al., 1979; Goldberg et al., 1991). Michel Goldberg and colleagues carried out a systematic investigation of the nature of aggregated material and its conditions of formation. When *Escherichia coli* tryptophanase was denatured in 8 M urea and subsequently renatured by dialysis, there was always a partitioning between active, renatured enzyme and aggregated material, depending on protein concentration

(London et al., 1974). If, after denaturation in 8 M urea, the protein was dialyzed against intermediate concentrations and finally dialyzed in urea-free buffer, a surprising low percentage of reactivation was observed in the range of concentrations around 3 M urea. Analytical ultracentrifugation of the material in the presence of 3 M urea revealed a polydisperse and highly aggregated material. When the same procedures were repeated in the presence of albumin or crude *E. coli* extracts that have not been induced for tryptophanase, the yields of renaturation versus aggregation were not affected. The ensemble of these results suggested that partially folded intermediates were populated in the refolding pathway and they were aggregating through specific intermolecular interactions (London et al., 1974). A model was proposed where partial unfolding/refolding could involve secondary structure segments that associate through native-like, but still incorrect, intermolecular associations with their complementary segments belonging to other molecules (London et al., 1974) (Fig. 1). The model was later extended to account for larger domains that form complementary

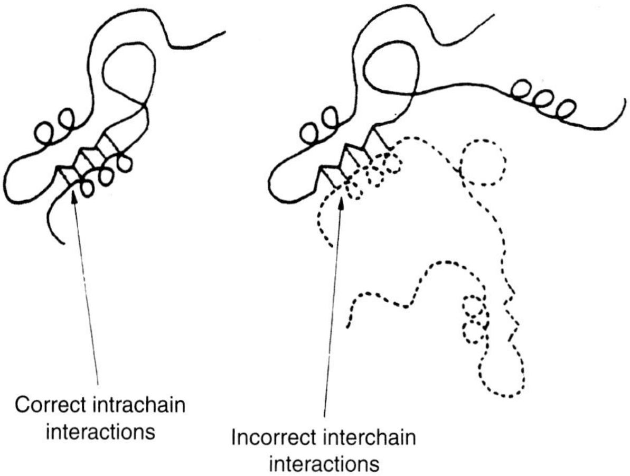

FIG. 1. Specific intermolecular interactions can mediate protein aggregation. Correct intramolecular interactions between segments of secondary structure are necessary for the formation of the native structure of the protein. Under conditions of partial unfolding, these interactions are weakened and intermolecular interactions formed between segments belonging to different chains mediate the aggregation process. The different chains are depicted in continuous and dotted lines. From London et al. (1974), reproduced with permission from Wiley.

interfaces within a native protein molecule; partial unfolding uncouples these domains and could lead to aggregation through interactions between complementary domains belonging to different protein chains (Goldberg and Zetina, 1980). This model is the intellectual precursor of the "domain swapping" hypothesis. "Domain swapping" occurs when partial unfolding leads to the "unhinging" of a domain or structural element termed "exchangeable" from the rest of a protein monomer (Fig. 2A). This "exchangeable" domain subsequently restores interactions with the "non-exchangeable" domain belonging to another protein monomer. Domain swapping was first experimentally observed in crystals of dimeric diphtheria toxin freeze-thawed from acid conditions and subsequently in other proteins (Bennett et al., 1994). These observations led to the hypothesis that "domain swapping" might have played a critical role in subunit exchanges and

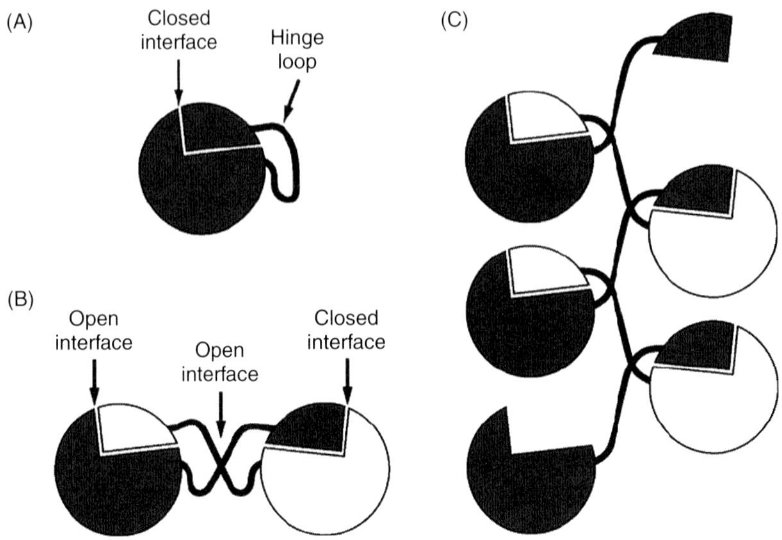

FIG. 2. The concept of "domain swapping." A native protein monomer is depicted as circle with an interface formed by two domains connected by a hinge (A). Partial unfolding leads to the "unhinging" of one domain from the rest of the monomer. This domain subsequently restores interactions with the other domain belonging to a different protein monomer, leading to creation of new interfaces and formation of a dimer (B). Swapping in an open-ended fashion can lead to formation of liner protein aggregates. From Nelson and Eisenberg (2006), reprinted with permission from Elsevier.

therefore the evolution of oligomeric protein states (Bennett et al., 1995) (Fig. 2B). By extrapolation, swapping of domains in an "open-ended" fashion has led to the important insight that such a mechanism could lead to protein aggregation (Liu and Eisenberg, 2002) (Fig. 2C).

III. Inclusion Bodies: Mechanisms of Formation and Structural Features

A common outcome during overexpression of heterologous proteins in *E. coli* is the accumulation of polypeptide chains in insoluble aggregates, termed "inclusion bodies." Inclusion bodies are refractile and can be distinguished in phase contrast microscopy and appear as electron-dense deposits in transmission electron micrographs, usually located at the poles of the cells (Taylor et al., 1986; Bowden et al., 1991). Upon cell lysis and centrifugation, inclusion bodies sediment at low speeds. Inclusion body formation was not due to the protein exceeding its solubility limits, as it usually happens in salting-out biochemistry procedures, since the process is not reversible by simple dilution of the chains. It soon became obvious that inclusion bodies were not consisting of covalently damaged polypeptide chains, since they could be solubilized under strong denaturing conditions and active protein recovered with various yields after elimination of denaturant (Marston, 1986; Kane and Hartley, 1988). Therefore, the information for reaching the native state was still present in the inclusion body chains, and their formation seemed to be due to folding problems within the cell.

A central difficulty in tackling the problem of inclusion body formation was the scarcity of model systems that could allow study of *in vivo* protein folding and comparison with folding *in vitro*. Such a comparison made possible a detailed study of the phage P22 tailspike folding and aggregation pathways by Jonathan King and colleagues. The unusual stability of the native form of this trimeric all-beta protein (resistance to Sodium Dodecyl Sulphate (SDS), temperature, and denaturants) made it possible to identify intermediates in the folding pathway that are sensitive to all these factors (Goldenberg and King, 1982; Goldenberg et al., 1983). The study of intracellular folding and aggregation revealed that aggregates were formed from early, soluble, partially folded intermediates in the folding pathway that could either proceed further to the native pathway (passing through a late "protrimer" intermediate having chains associated but not fully folded) or, alternatively, form intermolecular aggregates (Haase-Pettingell and King, 1988). This partitioning was highly temperature dependent, with the aggregation pathway favored at high temperatures. Furthermore, this partitioning is kinetically controlled, since chains synthesized at high temperatures can reenter the productive pathway upon shifting to low temperatures early enough, and alternatively, chains that reach the native state at low temperatures retain stability upon shifting to high temperatures (Betts and King, 1999; Betts et al., 2004). The folding and aggregation pathway is illustrated in Fig. 3. Several point mutations specifically affect the folding pathway, either by favoring the aggregation pathway (termed

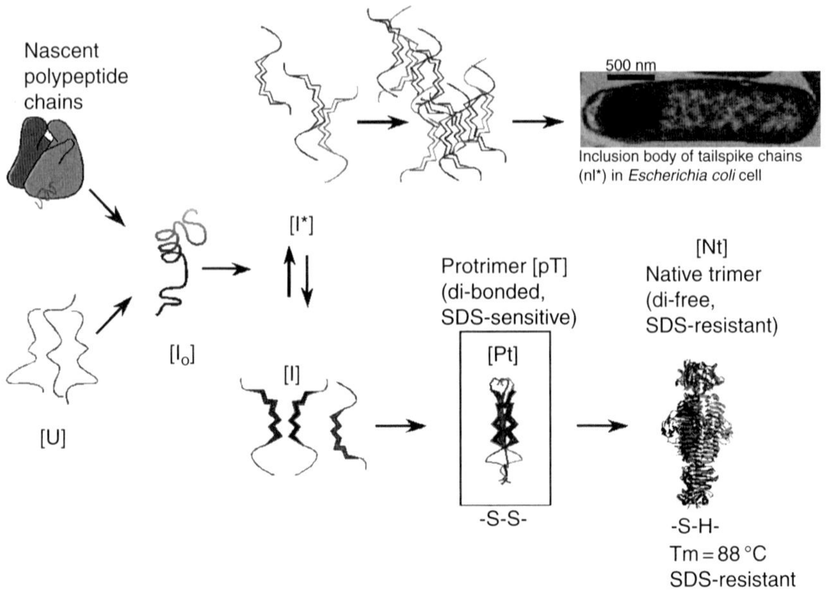

FIG. 3. Partitioning between folding and aggregation pathways for a model protein. The *in vivo* and *in vitro* folding and aggregation pathway of the trimeric, beta-structured P22 tailspike is depicted. After emergence from the ribosome or after dilution from denaturant *in vitro*, early partially folded intermediates can partition between the productive pathway (I) and the aggregation pathway (I*). In the productive pathway (bottom) they pass through an associated but yet not fully folded protrimer intermediate (pT) and following further conformation rearrangements they reach the native, highly resistant state (Nt). Alternatively, they associate and form intermolecular aggregates (nI*), visualized as inclusion bodies within bacterial cells (top panel). Point mutations specifically affect the portioning between productive folding and aggregation pathways. From Betts et al. (2004) reproduced with permission from Wiley.

"temperature-sensitive for folding" mutations) or by disfavoring the aggregation pathway (termed "second-site suppressors"). (Yu and King, 1984, 1988; Mitraki et al., 1991).

Extensive *in vitro* studies by Robert Seckler and colleagues and the information provided by the crystal structure of the protein subsequently made possible the thorough comparison of folding and aggregation pathways between *in vitro* and *in vivo* (Fuchs et al., 1991; Mitraki et al., 1993; Steinbacher et al., 1994; Seckler, 1998; Schuler et al., 1999) (Fig. 3). By using techniques of protein electrophoresis under native conditions, ladders of discrete oligomeric, nonnative folding intermediates in the *in vitro* folding pathway were trapped and clearly visualized (Speed et al., 1995). This methodology was found to be applicable for detection of intermediates in the refolding of other proteins, such as carbonic anhydrase and P22 coat protein (Speed et al., 1996; Betts et al., 1999). Subsequently two proteins with remarkably different sizes and distribution of folding intermediate ladders (tailspike and P22 coat protein) were separately denatured and allowed to refold in the same vessel, electrophoresed under native conditions, and probed by antibodies against each protein. No heterogeneous coaggregates of the two proteins were observed, indicating recognition specificity of partially folded intermediates in the aggregation pathway (Speed et al., 1996). Monoclonal anti-native and anti-denatured antibodies were used to probe the conformation of the productive and off-pathway intermediates. Both productive and off-pathway intermediates were not recognized by anti-native antibodies but were recognized by anti-denatured antibodies. This suggested that both classes of intermediates displayed common epitopes, probably corresponding to segments of the polypeptide chain that are involved in the aggregation process (Speed et al., 1997).

The detection of specific epitopes within folding intermediates, the failure to observe mixed aggregates, and the sensitivity of the aggregation process to point mutations suggested that aggregate formation was due to specific interactions between folding intermediates (Wetzel, 1994, 1996; Betts et al., 1997; Fink, 1998). When proteins fold *in vivo*, the presence of the native folding environment with the proper ions, cofactors, chaperone proteins, and so on is critical for the stability of the partially folded intermediates and their ability to enter the productive pathway. When expressed in heterologous hosts, the high concentration of the expressed chains may overwhelm the cellular machinery and in combination with the heterologous environment could favor the aggregation pathway for folding intermediates. The notion that inclusion bodies were formed as a result of specific interactions between folding intermediates implied that they were not "amorphous," but retained structural features that could be even "quasi-native" (Mitraki and King, 1989) (Fig. 4). However, the physical state of the aggregated material was

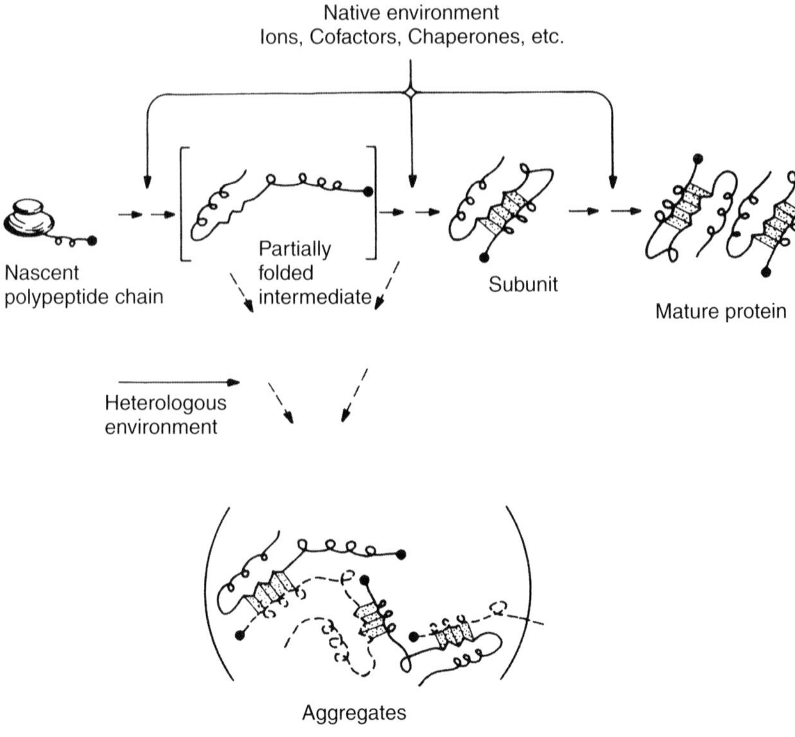

FIG. 4. General model for inclusion body formation. *In vivo* hypothetical folding pathway for a dimeric protein. Partially folded intermediates after release from the ribosome can enter the productive pathway and reach the native state. The native folding environment in the presence of the proper ions, chaperones, and so on is critical for proper folding. When expressed in heterologous hosts, the high concentration of the expressed chains may overwhelm the cellular machinery and in combination with the heterologous environment could lead to aggregation of the chains, always through specific, intermolecular interactions. From Mitraki and King (1989), reproduced with permission from Nature Biotechnology.

refractory to the use of protein structural characterization techniques that were developed for soluble proteins, with the exception of Fourier transform infrared (FTIR) spectroscopy (see below). The high-resolution structural characterization of the aggregated material was made possible only recently due to the availability of methods such as H/D exchange coupled with nuclear magnetic resonance (NMR) and solid-state NMR (see below).

IV. Protein Aggregation and Disease: Amyloids

Protein deposition in the form of insoluble aggregates composed of fibers in a cross-beta conformation is the hallmark of a family of diseases termed "amyloidoses" (Sipe, 1992; Sipe and Cohen, 2000). The most well known is Alzheimer's disease, a neurodegenerative disease characterized by extracellular deposition of a short, 40–42 amino acid peptide in plaques composed of "amyloid" fibers (Glenner, 1980a, 1980b). The consensus for the diagnostic traits of amyloid is the following: a) unbranched fibers with diameters around 100 Å as seen in electron microscopy (Serpell, 2000; Serpell and Smith, 2000), b) a cross-beta (i.e., beta-strands being perpendicular to the long fiber axis) X-ray fiber diffraction pattern with a meridional reflection of 4.7 Å corresponding to inter-strand spacing and a more diffuse equatorial reflection around 10 Å corresponding to beta-sheet spacing (Kirschner et al., 1986; Sunde and Blake, 1997; Sunde et al., 1997, Serpell et al., 1999), and c) binding of the dye Thioflavin T to give a characteristic shift in the dye's fluorescence emission spectrum, or Congo Red to give gold/green birefringence when viewed under crossed polarizers (Nilsson, 2004). Complementary to these are circular dichroism and FTIR spectra with peaks characteristic of intermolecular beta-sheet structure (Nilsson, 2004). Amyloid formation has also been associated with Huntington's disease (Perutz et al., 1994), Parkinson's disease (Conway et al., 2000), prion diseases (Prusiner et al., 1983), type 2 diabetes (Glenner et al., 1988), and with well-studied globular proteins, such as transthyretin (Colon and Kelly, 1992), beta2 microglobulin (Connors et al., 1985), and lysozyme (Booth et al., 1997). A significant body of *in vitro* folding studies of these latter systems with existing crystal structures suggested that amyloids form from partially folded intermediates through exposure of aggregation-prone regions

(Lai et al., 1996). It is generally accepted now that prefibrillar oligomeric aggregates are precursors of the fibrillar forms and these oligomeric forms seem to be associated with the toxicity, rather than the final amyloid fibrils (Come et al., 1993; Jarrett and Lansbury, 1993; Lansbury, 1997; Bucciantini et al., 2002; Caughey and Lansbury, 2003). Point mutations, often associated with familial forms of the diseases, dramatically influence the stability of folding intermediates and favor their aggregation pathways (McCutchen et al., 1993; Colon et al., 1996; Booth et al., 1997; Conway et al., 1998). For some mutant proteins, for example, the ones involved in familial lysozyme amyloidosis, crystal structures were solved and did not show any significant deviations from the wild-type global fold (Booth et al., 1997); however, they dramatically lower its stability against temperature and other denaturing agents. Many other proteins were subsequently found to be able to form amyloid fibrils under certain conditions; this surprising finding suggested that aggregate formation might be a general propensity of protein sequences (Chiti et al., 1999; Lansbury, 1999).

The structure of amyloid fibrils is a problem of vast complexity, given the variety of protein folds that can be accommodated within the fibril. Moreover, the size and insolubility of fibrils make this problem even more difficult to tackle. Short amyloidogenic peptides allowed study of this problem with X-ray fiber diffraction, cryo-electron microscopy, H/D exchange coupled with mass spectrometry, and solid-state NMR, and it led to the proposal of several structural models (Petkova et al., 2002; Jaroniec et al., 2004; Luhrs et al., 2005; Makin et al., 2005; Paravastu et al., 2008). A common feature for all these models is a backbone structure formed by single beta strands in either parallel or antiparallel conformation. The single-stranded model can be extended to accommodate hairpin building blocks where each peptide contributes a beta-strand-turn-stand motif to the fibril architecture.

David Eisenberg and colleagues recently obtained high-resolution structural information for fibril-forming peptides from the sup35 yeast amyloid (see below), Alzheimer's peptide, tau, and amylin (Nelson et al., 2005; Sawaya et al., 2007). The ability to grow microcrystals from these peptides (Diaz-Avalos et al., 2003) and the use of the microfocus beamline at the European Synchrotron Radiation Facility (ESRF) contributed to this remarkable achievement (Nelson et al., 2005). The relevance of these needle-shaped microcrystals to amyloid fibrils is corroborated by the fact that they often coexist in solutions, with fibrils seen to emanate from the tips of the crystals. Furthermore, the simulated X-ray fiber diffraction patterns of these peptide microcrystals are in good agreement with the X-ray diffraction patterns of fibrils formed by entire proteins; the microcrystals can also seed fibril formation from the full-length proteins. The first structure of the sequence GNNQQNY from sup35 revealed a pair of beta-sheets with strands running perpendicular to the long axis of the fibril; side chains are tightly interdigitated to form a dry interface, termed the "steric zipper." The "steric zipper" interface is repeated along the length of the long axis of the microcrystal. Eight classes of these steric zippers were subsequently reported from eleven new peptides (Sawaya et al., 2007). The "cross-beta spine" models are particularly attractive since they allow the accommodation of the remaining protein in the form of natively folded domains that decorate the periphery of the spine (Nelson and Eisenberg, 2006) (Fig. 5A). Furthermore, they can accommodate domain-swapped units that combine at the edges of the beta-spine (Fig. 5B). The relevance of these two possibilities will be further discussed below.

V. Fusion Proteins in the Study of Inclusion Bodies and Amyloids

Given the experimental difficulties in applying methods that could specifically follow the intracellular fate of expressed proteins, fusion proteins have been used extensively as reporters of folding and aggregation for target proteins/sequences. In the first reported studies, fusion proteins were combined with aggregation-prone sequences in arrangements that allowed the reporter protein to fold and acquire biological activity while the target protein/sequence was able to fold into an intracellular soluble form (Waldo et al., 1999; Wigley et al., 2001; Wurth et al., 2002). Alternatively, when the target protein/sequence was aggregating, the reporter protein would misfold and be incorporated in aggregates, resulting in loss of biological activity. Proteins with enzymic activity, such as beta-galactosidase, and fluorescent proteins such as green fluorescent protein (GFP) were successfully used in these approaches (Waldo et al., 1999; Wigley et al., 2001; Wurth et al., 2002). As GFP fluorescence is acquired only if the polypeptide backbone is properly folded in order to allow cyclization and oxidation of an internal Ser–Tyr–Gly tripeptide

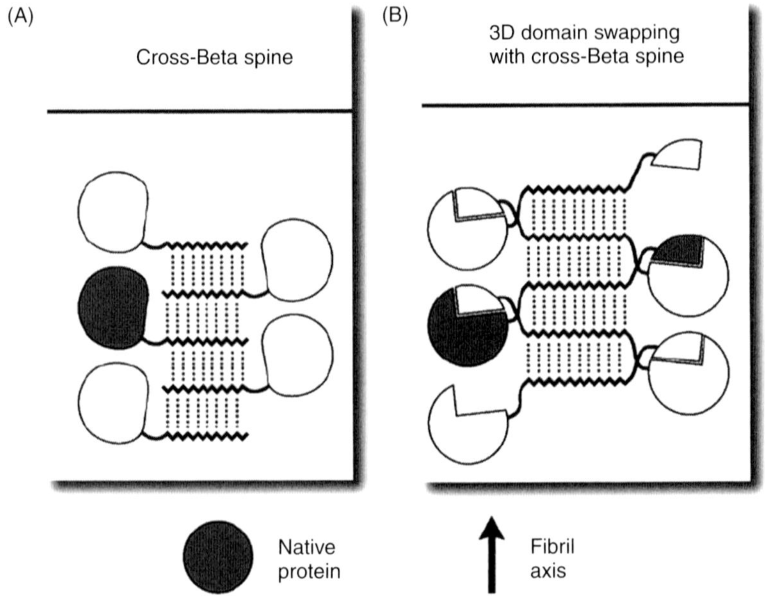

FIG. 5. The cross-beta spine model and accommodation of native domains at the periphery of the spine. An amyloid-forming segment or even a short sequence stretch can "unhinge" and form the cross-beta "spine" of an amyloid fibril. Beta-strands are represented here with the zig-zag line and hydrogen bonds with the dotted lines. The rest of the domain is accommodated at the periphery of the spine and can retain native-like structure and activity (A). (B) The combination of a beta-spine with swapping domains. From Nelson and Eisenberg (2006), reprinted with permission from Elsevier.

motif, acquisition of the chromophore is considered to reflect the proper folding of the protein into its native state. As more studies gradually appeared in this area, enzymatic activity/fluorescence was reported, surprisingly associated with the inclusion body fraction of bacteria, in the cytoplasm as well as in the periplasmic space. When beta-galactosidase or GFP were C-terminally fused to an aggregation-prone sequence such as the 23-kDa VP1 capsid protein from the foot-and-mouth disease virus, fluorescence was colocalized within the inclusion body areas observed in the phase contrast microscope. Enzymatic activity of beta-galactosidase was also found to be associated with the inclusion body fraction (Garcia-Fruitos et al., 2005). With a similar approach designed to monitor protein folding and aggregation in the periplasmic space, the aggregation-prone variant MalE31 of the maltose-binding protein (Mal E) (Betton and Hofnung, 1996; Saul et al., 2003) was fused to the N-terminus of beta-lactamase or alkaline phosphatase (phoA) in the presence of small oligopeptide linkers between them (Arie et al., 2006). The fusion proteins accumulated in insoluble form in the periplasm, but catalytic activity of the fusion moieties was maintained. The proteins in the insoluble fraction were not degraded by intracellular proteases, as they were recognized by both anti-MalE31 antibodies and antibodies against fusion proteins. On the contrary, constructs comprising the wild-type soluble MalE protein upstream of the fusion proteins were accumulated in soluble form in the periplasm and were partially degraded intracellularly, probably through cleavage of the linker region, releasing the fusion counterparts in active form. Since MalE31 fusion proteins were found to be resistant to degradation, the possibility that partial release of the fusion counterparts would result in soluble activity simply being associated with an insoluble, inactive form was excluded. The intracellular localization of the catalytic activity of the phoA was further confirmed by using a fluorescent substrate; the fluorescence was punctually localized in small foci. When using wild-type MalE fused to phoA, uniform distribution of fluorescence was observed indicating free diffusion of the fusion protein in the periplasm (Arie et al., 2006). To further investigate the relevance of this observation to the proteins *per se* in the absence of fusion counterparts, the authors studied fusions that joined the wild-type and aggregation-prone variants of MalE in the presence of small oligopeptide linkers. When the soluble wild-type (MalE) was present upstream of the insoluble variant (MalE31), the fusion proteins were produced in soluble periplasmic form; when MalE31 was present upstream of MalE, the fusion proteins were produced in aggregated form. It would therefore seem that the intrinsic folding propensity of the upstream proteins determine the intracellular fate of the fusion proteins. To interpret these findings was intellectually challenging. It was reasonable to hypothesize that the folding characteristics of the upstream domain were a dominant force in driving aggregation, but the downstream domains could still assume/maintain their native fold and activity, even while being physically sequestered in an insoluble aggregate (Arie et al., 2006).

Further evidence for this hypothesis arises from studies of fusion proteins linked to amyloid-forming sequences, both *in vitro* and *in vivo*. Such fusions were first used to monitor intracellular fates of yeast prion proteins (Edskes et al., 1999). In an attempt to elucidate the architecture of the Ure2p prion filaments, Baxa et al. (2003) fused the N-terminal prion domain of the Ure2p protein (see below) to barnase, carbonic anhydrase, glutathione transferase, and GFP. The fusion proteins form insoluble filaments that clearly comprise a central filamentous core surrounded by globular appendages, as seen both by cryo-EM imaging and by scanning transmission electron microscopy (STEM) after vanadate staining. In no case did the globular fusion proteins form fibrils under the same conditions. The green fluorescence of GFP was associated with the insoluble fibrils and was released into the soluble fraction after protease treatment. The diameters of the fusion protein filaments (14–25 nm) were reduced to 4 nm after proteolysis, in agreement with the fibril diameters of the prion domain alone. This study provided compelling evidence for an amyloid-forming domain driving aggregation into fibrils, forming a central "spine" with attached globular moieties protruding from the spine (Baxa et al., 2003) (Fig. 5A).

VI. Short Sequences may Suffice to Drive Aggregation

Are short, amyloid-forming sequences sufficient to drive aggregation when fused to larger globular domains and thereby "tether" the globular counterparts in biologically active form on the amyloid aggregate? How could this affect/challenge the well-established notion of specificity of interactions that drive protein aggregation? Fluorescence resonance energy transfer (FRET) techniques were previously developed to probe the specificity of aggregation of unrelated, aggregation-prone proteins within inclusion bodies ("aggresomes") in eukaryotic cells (Rajan et al., 2001). Fluorescence energy transfer occurs between two fluorophores when they are located in distances ranging within 10–100 Å from each other and is therefore a sensitive method to assess proximity of fluorescent proteins. When representative aggregation-prone proteins labeled with different fluorescent protein tags were coexpressed in the same cell, Kopito and colleagues demonstrated, using FRET and fluorescence microscopy, that these proteins do not coaggregate (Rajan et al., 2001). Digital deconvolution microscopy further showed that each protein was sequestered within homogenous discrete foci even when the different proteins seemed to colocalize within a large aggregate. This approach was extended to address the question of whether protein stretches that share homologous/complementary aggregation sequences could coaggregate. For example, long polyglutamine stretches are associated with huntingtin aggregation, while short ones remain soluble and diffusely cytoplasmic when expressed in cells (Kazantsev et al., 1999). When long polyglutamine stretches (Q103) fused to cyan fluorescent protein (CFP) were coexpressed with short stretches (Q25) fused to yellow fluorescent protein (YFP), a strong FRET signal was observed. This experiment suggested "recruitment" of the short poly Q constructs by the long ones within the aggregate, with the fusion proteins remaining active. Ventura and colleagues used FRET techniques to address the above questions with the Alzheimer's 42 amino acid peptide (Aβ42) fused to different fluorescent proteins (BFP, blue fluorescent protein, and GFP) and coexpressed in the same cell (Morell et al., 2008). The same approach was used for the VP1 capsid protein from the foot-and-mouth disease virus. The FRET observed when the two different fluorescent proteins were coexpressed with the same aggregation-driving domain (i.e., either Aβ42 or VP1) was three-fold higher than the FRET observed between proteins bearing different aggregation domains. Strikingly, inclusion bodies of the cells coexpressing Aβ42–BFP and VP1–GFP presented a segmented morphology, with the green fluorescence signal concentrated at the poles of the inclusion bodies, while the blue fluorescence being uniformly distributed in the inclusion bodies (Morell et al., 2008). Therefore the relative segregation of the different aggregating sequences within the same inclusion body reinforced the notion of the specificity of interactions driving inclusion body formation.

The sequences that drive aggregation when fused/inserted in globular moieties can be very short. "Transplanting" short amyloid-prone sequences within nonamyloidogenic proteins was reported to induce amyloid formation (Ivanova et al., 2004; Ventura et al., 2004). Further experimental evidence was provided by the insertion of a short glutamine-rich sequence flanked by two glycines (G Q10 G) in a hinge loop of the RNase A molecule (Sambashivan et al., 2005). This hinge loop, residues 112–115, connects the N-terminal domain (residues 1–111) and a carboxyl-terminal segment (residues 116–124) and is involved in domain swapping when RNase is freeze-dried from 40% acetic acid (Liu et al., 2001). When the hinge loop was expanded with the glutamine-rich sequence, amyloid fibrils formed. In RNase A, two histidine residues critical for activity are located in each one of the domains (Crestfield and Moore, 1963). When RNase A molecules bearing two different mutations, each one mutating a His to Ala within each domain, are mixed and subsequently freeze-dried, the fibrils formed show significant RNase activity. These results are compatible with a model where the aggregating sequence forms a central cross-beta spine with the globular domains protruding and still be able to complement to form an active molecule (Fig. 5B). It is interesting to note that a linker consisting of a nine glycine residue expansion did not form fibrils. This approach was further expanded by introducing in the linker region 6- to 8-residue-long amyloidogenic sequences from tau, Alzheimer's peptide, prion protein, and alpha-synuclein (Teng and Eisenberg, 2009). They all were found to induce amyloid fibril formation. This observation reinforces the argument that exposure of short aggregation sequences, or "fibril-forcing segments," could be sufficient to induce aggregation of otherwise normally soluble proteins.

The body of evidence emerging from these studies suggests that it is possible for an aggregation-prone domain, or even a short aggregating sequence, to "recruit" a fused globular counterpart within aggregates, with the latter able to fold into an active conformation. The question arises, though, whether this could be observed in a "real-life" situation, within, for example, the sequence of the same, natural protein. A study by Tokatlidis et al. reported an exceptional and puzzling result at the time it was published: the existence of inclusion bodies that retained biological activity, implying "native-like" conformation of the sequestered protein. When expressed in *E. coli*, endoglucanase EGD from *Clostridium thermocellum*, a 68-kDa protein involved in cellulose degradation was found in inclusion bodies (Tokatlidis et al., 1991). Using a small diffusible substrate, significant biological activity (close to the specific activity of the soluble enzyme) was found in the inclusion body fraction *in situ* and was confirmed to remain in this fraction and not to be released in soluble form under the assay conditions. Upon solubilization of inclusion bodies with urea, a 65-kDa protein was released as a result of cleavage by an endogenous protease. The 65-KDa protein corresponds to the catalytic N-terminal domain of the protein, whereas the polypeptide excised by the protease C-terminal domain consists of two homologous segments of 23 amino acids each. This C-terminal domain is devoid of catalytic activity and is highly conserved in endoglucanases and xylanases. The biological role of this C-terminal segment consists in tethering the catalytic domains to the cellulosome, a macromolecular complex involved in cellulose degradation, and is referred to as the "dockerin domain" (Tokatlidis et al., 1991). Furthermore, grafting the dockerin to an endoglucanase that does not possess such a domain (endoglucanase C) attaches this subunit to the cellulosome (Tokatlidis et al., 1993). When the 65-kDa catalytic domain alone is expressed, it is produced in soluble and active form. The puzzling result of catalytic activity present in an insoluble fraction of inclusion bodies was insightfully interpreted at the time: inclusion body formation would be mediated by intermolecular interactions involving the C-terminal domain, recruiting the catalytic domain in its active form (Tokatlidis et al., 1991). In light of the body of evidence above, furthermore, this interpretation makes sense. In addition, it opens the possibility that similar mechanisms could be operating in proteins that comprise segments/domains with dramatically different aggregation propensities. The relative rates of folding of the two domains and the rate of aggregation/recruitment should play a decisive role in this process, as well as the length and nature of the sequences that link them. We will also discuss below potential applications in the biotechnological production of proteins.

VII. Amyloid Character of Inclusion Bodies

As noted above, inclusion body formation and amyloid formation share two important features that suggest sequence-specific aggregation: oligomeric intermediates as precursors to the aggregated sate and sensitivity to site-specific point mutations. Inclusion bodies were reported to present amyloid-type characteristics, such as binding to Thioflavin T and Congo Red, intermolecular beta-structure as seen by FTIR, and ability to seed, after extracted from cells, *in vitro* amyloid fibril formation of homologous soluble counterparts (Carrio et al., 2005). Earlier FTIR studies revealed in many instances an increase in intermolecular beta-sheet structure, or alternatively, persistence of native-like FTIR spectra in the inclusion body fraction (Oberg et al., 1994; Przybycien et al., 1994; Fink, 1998). Recent studies reported the possibility of recording FTIR spectra in intact cells and of following the kinetics of inclusion body formation in real time (Ami et al., 2005, 2006). However, these techniques yield averaged information and do not allow sequence-specific structural information to be extracted.

Inclusion bodies were made amenable to detailed structural characterization only recently, by using H/D exchange in inclusion bodies pelleted after cell lysis (Wang et al., 2008). Inclusion bodies were resuspended in D_2O to allow H/D exchange, followed by quenching, solubilization, and NMR peak assignment. Loss of peak intensity is indicative of fast exchange whereas persistence of peaks identifies solvent-protected backbone amide protons already involved in hydrogen bonds. Three inclusion body-forming proteins representing different structural groups (i.e., alpha helical, beta-sheet, or alpha/beta mixed structures, with or without disulfide bonds) were studied. These proteins were a) the alpha-helical, 6-kDa early secreted antigen (ESAT-6) from *Mycobacterium tuberculosis*, which is able to fold only in the presence of another protein partner (CFP 10) both *in vitro* and during *E. coli* overexpression, b) the mixed alpha-helical/beta-sheet fragment corresponding to residues 13–74 of the human bone morphogenetic protein (BMP-2), which cannot fold *in E. coli* due to the absence of the residues 75–114 and the inability to form disulfide bonds in the *E. coli* cytoplasm, and c) the all-beta extracellular domain of the human myelin oligodendrocyte protein, MOG (ECD), which forms inclusion bodies in *E. coli*, presumably due to an inability to form its disulfide bond. For all three proteins, most of the residues were found to be weakly or not protected, and therefore were assumed to adopt a disordered conformation. Short (7–10 amino acids) stretches were found to be protected from hydrogen-exchange segments. X-ray fiber diffraction of inclusion bodies showed the characteristic amyloid signature, suggesting that these segments could adopt a cross-beta structure. The amyloid character was confirmed with ThT binding and Congo Red binding with green birefringence under crossed polarizers. Strikingly, electron micrographs of freshly extracted inclusion bodies show electron-dense, round objects where small fibrillar morphologies can occasionally be seen. After *in vitro* incubation in buffers for some hours, these fibrillar structures become more evident (Fig. 6). Complementary approaches were used to confirm that the protected segments were indeed critical in driving inclusion body formation. Synthetic peptides corresponding to these segments form amyloid fibers *in vitro*, even when they do not adopt beta-structure within the native protein context. When residues belonging to these segments were mutated to Arg (a charged amyloid-breaking residue), the expressed protein bands were either moved to the soluble fraction or disappeared, possibly due to intracellular degradation. Mutating residues belonging to

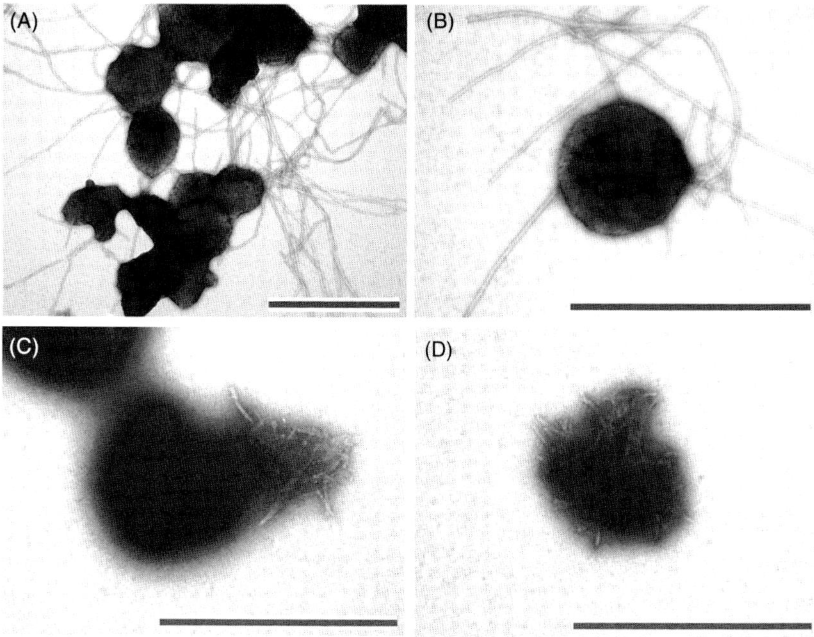

FIG. 6. Amyloid fibrils associated with inclusion bodies. (A, B) Transmission electron micrographs from purified inclusion bodies incubated *in vitro* for 12 hours at 37 °C. (C, D) Freshly purified inclusion bodies of BMP2 (13–74) from *E. coli* cells, after overexpression for 12 hours at 37 °C. In both cases, fibrils emanating from the inclusion bodies are observed. Scale bars indicate 1 micron. From Wang et al., 2008, PLoS Biol 6(8): e195. doi:10.1371/journal.pbio.0060195.

the rest of protein did not have any effect on inclusion body formation. Deleting the aggregation-prone segment of the ESAT-6 protein or replacing it with a nonaggregating segment abolished inclusion body formation. Finally, fusion of the aggregation-prone segment of BMP to a protein that is highly soluble when expressed in *E. coli* resulted in inclusion body formation of the fusion construct. These results suggest the existence of short segments within a variety of protein structures that could be involved in intermolecular beta-structure formation and mediate amyloid and inclusion body formation (Wang et al., 2008). More recently, solid-state NMR of the inclusion bodies formed from the prion domain of the Het-s protein (see below) from the filamentous fungus *Podospora anserina* suggested that they share the same amyloid structure with amyloid filaments formed by the same domain *in vitro* (Wasmer et al., 2009).

In another recent study, solid-state NMR of whole cells was used to probe the conformation of specific residues within inclusion bodies formed by influenza virus hemagglutinin expressed in *E. coli* (Curtis-Fisk et al., 2008). The bacteria were grown in minimal media containing ^{13}C- and ^{15}N-labeled amino acids and the method used allowed assessment of backbone ^{13}CO groups bonded to ^{15}N. The authors reported native-like, alpha-helical conformations for four selected residues (Curtis-Fisk et al., 2008). Reports from the above two studies are not necessarily conflicting, since the involvement of short stretches in beta-amyloid structures does not preclude the adoption of native-like conformations by the remaining polypeptide chain, or parts of it.

VIII. Natural Amyloids

Fibrous amyloid-structured aggregates are not only involved in misfolding and disease, but can also assume the role of structural and functional nanobiomaterials (Fowler et al., 2007). Yeast and fungi proteins can play such potential functional roles. The yeast protein sup35 is involved in the regulation of stop-codon reading, the yeast protein Ure2p in the regulation of nitrogen catabolism, and the Het-s protein from the fungus *P. anserina* in the regulation of heterocaryon formation. Each of these proteins can exist in either soluble or amyloid form. The formation of cytoplasmic amyloid results in phenotypes that are inherited in a non-Mendelian, epigenetic manner. In these cases, amyloids are thought of as self-replicative protein structures that can function as molecular "memories" for the transmission of genetic information, and they could confer evolutionary advantages. Yeast prions are a vast and complex domain of amyloids, and reviewing them in detail is beyond the scope of this chapter. The reader is referred to pertinent reviews (Bousset and Melki, 2002; Shorter and Lindquist, 2005; Baxa et al., 2006; Saupe, 2007; Wickner et al., 2007).

Functional amyloids have also recently been reported for mammals. A fragment of the protein Pmel forms amyloid fibrils in organelles called the melanosomes, and these fibrils serve as templates for the chemical synthesis of melanin (Fowler et al., 2006). Their functional role is supposed to be protective, since they seem to sequester highly cytotoxic melanin precursors and prevent their leakage out of the melanosome. Peptide and protein hormones seem to be stored in secretory granules in an amyloid form (Maji et al., 2009). This guarantees both a long-term storage in a stable depot form and controlled release of monomers upon signaling.

Nature uses amyloids as materials whenever strength and protection is needed or as adhesion materials. One of the first cross-beta structure patterns that was reported by X-ray diffraction was the pattern of protein fibrous stalks of the green lacewing flies *Chrysopa flava* (Geddes et al., 1968). These fibers are tethered on the surface of plant leaves and subsequently the eggs are laid at the tip of the fibrous stalks. The stalks seem to protect eggs against predation and cannibalism (Ruzicka, 1997). A classical example is also insect chorion (eggshell) proteins, extracellular proteinaceous formations surrounding the oocytes in silk moths. The chorions have extraordinary mechanical and chemical stability, protecting the developing oocyte from environmental hazards such as temperature, pressure, bacteria, and viruses. X-ray diffraction studies on intact chorions reported the characteristic reflections of cross-beta structure (Hamodrakas et al., 1983). The family of chorion proteins comprise a central highly conserved domain with glycine-rich amino acid repeats flanked by N- and C-terminal variable domains. Synthetic peptides corresponding to a 51-residue consensus sequence of the central domain and a shorter, 18-residue peptide self-assemble into amyloid-type fibrils *in vitro* as indicated by X-ray fiber diffraction, FTIR and Raman spectroscopy, and Congo Red binding (Iconomidou et al., 2000, 2001). These studies further supported the notion that chorions are natural protective amyloids (Iconomidou and Hamodrakas, 2008). Egg envelopes in the fish *Austrofundulus limnaeus* were also found to be composed of protein fibrils with diameters ranging from 4–6 nm that stain with Congo Red and display green birefringence. The chorion proteins are mostly beta-structured with a high turn content (as seen by FTIR) and assemble into fibrils also *in vitro*, after denaturation in 8-M guanidine hydrochloride and subsequent removal of the denaturant by dialysis. The protective role in this case consists in preventing dehydration under conditions of water stress (Podrabsky et al., 2001).

Another remarkable class of natural amyloids is formed by fungal proteins called hydrophobins (Gebbink et al., 2005; Linder et al., 2005;

Sunde et al., 2008). Hydrophobins are secreted by fungi and self-assemble at the air–water interface forming amphipathic, SDS-resistant membranes. For a class of hydrophobins, called class I hydrophobins, the self-assembled membranes have a characteristic morphology, namely, they consist of rodlets with amyloid signatures, such as Congo Red binding/green birefringence and the characteristic X-ray fiber diffraction pattern (Butko et al., 2001; Gebbink et al., 2005; Linder et al., 2005; Sunde et al., 2008). The biological role in this case consists in forming an amphipathic coating that covers the fungal hyphae, allowing them to breach the air–water interface, and subsequently rendering the aerial hyphae hydrophobic and facilitating their dispersion in air. In this way the attachment to plant or animal surfaces and their subsequent colonization is facilitated. Gram-positive bacteria such as streptomycetes use similar proteins, called "chaplins," that form similar amphipathic membranes and allow their spores to escape damp environments and adhere to surfaces and tissues (Claessen et al., 2003). Curli fibers are assembled at the surface of enteric bacteria such as *E. coli*; they play important roles in attachment and biofilm formation and are one class of extensively studied natural amyloids (Barnhart and Chapman, 2006). A major protein subunit called CgsA is secreted in soluble, unstructured form and is subsequently polymerized into amyloid-type fibrils in the presence of a minor "nucleator" protein, (CgsB) anchored in the membrane (Chapman et al., 2002; Hammer et al., 2007; Wang et al., 2007). Five imperfect sequence repeats rich in Asn, Gln, and Gly comprise the protease-resistant, amyloid core of CgsA fibrils (Wang and Chapman, 2008). The curli system allows combination of *in vitro* and *in vivo* approaches with mutagenesis and synthetic peptide approaches and is therefore a very good model system for the study of natural amyloids.

A number of recent papers report involvement of amyloid-type nanostructures in adhesion processes. Bacteria-producing amyloid-type adhesins were identified in several phyla from biofilms of four different natural habitats (Larsen et al., 2007). Amyloid-type nanostructures were also detected in unicellular subaerial algae, multicellular subaerial algae, and diatoms (Fukuma et al., 2006; Mostaert et al., 2006, 2009). All of the above studies used the Thioflavin T and Congo Red tinctorial assays in combination with amyloid-specific antibodies for the biofilms in natural habitats. X-ray fiber diffraction patterns will be needed for confirming the amyloid character of these extracellular adhesives. However, it is understandable that this represents a technical challenge for these specimens, and further developments will be needed to make it possible. It is notable that atomic force microscopy (AFM) force–extension curves from these natural adhesives present a characteristic "sawtooth" mechanical response (Mostaert et al., 2006), previously also reported for natural adhesives, fibers, and composites (Smith et al., 1999). This type of mechanical response could correspond to unraveling of "sacrificial bonds" and "hidden length" in the material, characteristics that are designed to confer mechanical benefit (Fukuma et al., 2006).

IX. Protein Biomaterials: Amyloids as Bio-Nanomaterials

Fibrous nanostructured objects, such as carbon nanotubes, have attracted great interest recently for their potential integration in future generations of nano-, micro-, and macrodevices with possible industrial applications (Sinnott and Andrews, 2001; Terrones, 2003). Protein-based nanofibers present an attractive alternative to inorganic nanofibers, since they can assemble under mild aqueous conditions, are considerably less expensive to manufacture, and are more easily configurable, that is, they allow the attachment or engineering of additional functionalities. Sequence repeats are a common feature of amyloid-forming proteins and assume the role of self-assembling building blocks (Gilead and Gazit, 2005; Gazit, 2007a). The controlled assembly from building blocks and the ability to introduce modifications at the sequence level make them ideal scaffolds for the rational design of nanobiomaterials. There is, therefore, a clear advantage for the development of new, protein- and peptide-based nanofibers and nanotubes with the ultimate aim of integrating them in nanotechnological assemblies and devices. The exceptional mechanical properties of amyloid fibers and their chemical resistance to extreme conditions such as heat, temperature, denaturants, and detergents make them particularly attractive for interfacing with the world of "hard" inorganic materials.

Two pioneering papers that were the first to create a material concept for amyloids appeared in 2003. Scheibel and Lindquist introduced cysteine residues at the sequence of the sup35 yeast prion protein; this led to templating of metal nanoparticles and the fabrication of conducting nanowires (Scheibel et al., 2003). Gazit and colleagues discovered that the dipeptide Phe–Phe, a core recognition motif of the Alzheimer's peptide sequence, forms hollow nanotubes upon self-assembly *in vitro*. These hollow nanotubes were subsequently used as a template for casting metal nanowires. After digestion of the template with proteinase K, metal nanowires with a diameter of 20 nm were fabricated (Reches and Gazit, 2003). Peptides from natural fibrous proteins can also self-assemble into amyloid fibrils (Papanikolopoulou et al., 2005) and can be designed for binding of metal nanoparticles (Kasotakis et al., 2009) (Fig. 7).

In the last decade an increased understanding of the mechanisms governing protein and peptide self-organization has encouraged distinct research directions, ranging from biomedicine and biotechnology to materials science and bio-nanotechnology (Zhang, 2003; Rajagopal and Schneider, 2004; Gazit, 2007b). Peptide nanotubes were used as recognition and surface-enhancing elements in bio-electrochemical sensor applications (Yemini et al., 2005a, 2005b). The formation of hydrogels under certain conditions is also a distinctive property of self-assembling peptides (Aggeli et al., 1997). Such self-supporting hydrogels are increasingly being used as scaffolds for applications in tissue engineering and drug delivery (Gelain et al., 2007). Hydrogel formation can also be induced by

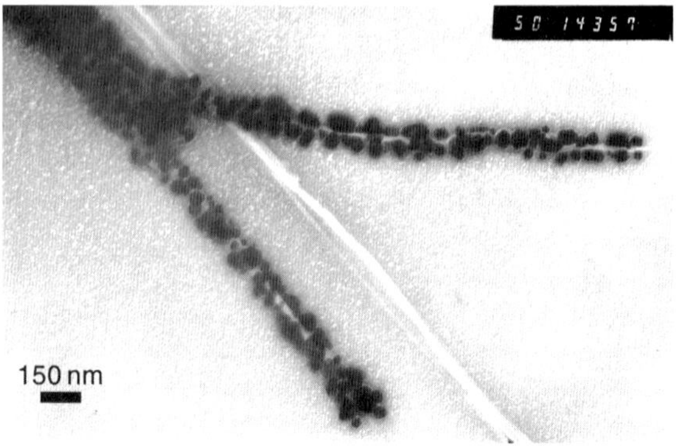

FIG. 7. Amyloid fibrils as templates for inorganic nanomaterials. Amyloid fibrils designed to display metal-binding amino acids were incubated with gold tetrachloroaurate and subsequently reduced with sodium citrate. They were imaged in transmission electron microscope after negative staining with uranyl acetate. In this image, a "naked" fibril is shown along with a fibril decorated with gold nanoparticles. From Kasotakis et al. (2009), reproduced with permission from Wiley.

self-assembly of peptides via beta-strand swapping. Schneider and colleagues designed beta-hairpin peptides alternating hydrophobic residues with positively charged ones that adopt a random coil structure at room temperature. The design includes an "overhanging" sequence at the N- or C-terminus of the hairpin that could function as the "exchangeable" element capable of strand swapping (Fig. 8). Upon heating, swapped hairpin dimers form and further self-assemble into fibrils, forming hydrogel networks (Nagarkar et al., 2010). The length and composition of the "exchangeable" strand seems to influence fibril morphology as well as bulk rheological properties. This work demonstrated that swapping is possible for small peptides and opens the way for rational design of self-assembling biomaterials with tunable properties. A particularly exciting possibility offered by designing self-assembling biomaterials based on

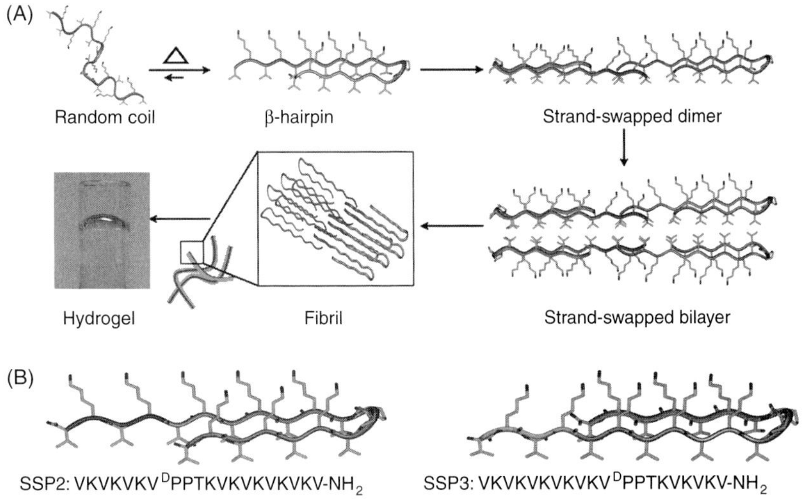

FIG. 8. "Domain swapping" in materials design. (A) A designed peptide has a random coil structure, folds into a beta-hairpin upon heating, and subsequently self-assembles and forms hydrogels. (B) Schematic drawing of the designed peptides. The blue part depicts an "exchangeable" sequence that is capable for "strand swapping" located at the N-terminus of the peptide. The cyan part depicts an "exchangeable" sequence that is capable for "strand swapping" located at the C-terminus of the peptide. The "nonexchangeable" domain is depicted in orange. The sequences of the two peptides in one-letter code are shown at the bottom. From Nagarkar et al. (2010), reprinted with permission from Wiley (See color plate 6).

strand and domain swapping would be the controlled display of functional units along the self-assembled structural material "spine."

As we mentioned above, protein and peptide hormones seem to be stored in amyloid form within secretory granules (Maji et al., 2009). This observation, in combination with the observation that long-acting gonadotropin hormone analogs form gels (a frequent indication of self-assembly and amyloid formation), stimulated recently the concept of amyloid as a source of long-term storage and controlled delivery of protein and peptide drugs (Maji et al., 2008). A protein or peptide drug that is able to form fibrils immediately or shortly after administration can ensure the sequestration of the active monomer in a stable depot and its subsequent gradual release from the fibril termini, resulting in sustained and controlled delivery of the active principle. However, safety concerns have to be addressed before such pharmaceutical approaches can be applied.

Inclusion bodies are considered to be a major bottleneck in industrial protein production. An understanding of their formation mechanisms led to proposals of several strategies to avoid their formation and obtain proteins in soluble and active form inside bacteria (Mitraki and King, 1989; Schein, 1989; Georgiou and Valax, 1996; De Bernardez Clark et al., 1999; Baneyx and Mujacic, 2004; Rinas et al., 2007). However, in light of recent structural insights, reviewed in this chapter, alternative production strategies could be envisaged. For example, fusion of catalytic moieties to short aggregating sequences with use of appropriate linker sequences could result in formation of inclusion bodies where the catalytic moiety can fold into its active form. This would offer a purification advantage with the possibility of subsequently releasing the active protein from the aggregated "spine" by selective proteolysis. Finally, the use of inclusion bodies as particulate, hydrated, and porous materials was recently envisioned. Inclusion bodies were patterned on amino-terminated silicon surfaces and were shown to support cell attachment and proliferation (Garcia-Fruitos et al., 2009).

X. Summary and Outlook

In the last 30 years, the field of protein aggregation, inclusion body formation and amyloid formation witnessed numerous unexpected conceptual turns. Currently, aggregation and inclusion body formation are processes considered worth analysis and the subject has come under intense investigation. The concept of natural and functional amyloid has emerged, and subsequently the notion of exploiting amyloid structures for beneficial purposes. More recently, an unexpected structural correlation between inclusion bodies and amyloids has been reported, though it remains to be confirmed by the investigation of more systems. In this chapter, an attempt is made to summarize the underlying common principles that govern structure and formation of inclusion bodies and amyloids. In the years to come, we should expect to gain more structural insight as a consequence of high-resolution methods for studying protein aggregates. Furthermore, reports of more "cross-seeding" or "swapping" between the domains of structural biology and natural biomaterials are highly probable. In the concluding paragraph of Anson and Mirsky's (1931) paper, a "biological significance of coagulation" was foreseen. There they stated that "it is difficult to believe that Nature has not exploited the coagulability of proteins".

Acknowledgments

The author thanks Dr. Kostas Tokatlidis for useful input and Emmanouil Kasotakis for help with the figures. Funding from the European Union (STREP NMP-CT-2006-033256, "BeNatural") is gratefully acknowledged.

References

Aggeli, A., Bell, M., Boden, N., Keen, J. N., Knowles, P. F., McLeish, T. C. B, Pitkeathly, M., Radford, S. E., (1997). Responsive gels formed by the spontaneous self-assembly of peptides into polymeric beta-sheet tapes. *Nature,* **386**, 259–262.

Ami, D., Natalello, A., Gatti-Lafranconi, P., Lotti, M., Doglia, S. M., (2005). Kinetics of inclusion body formation studied in intact cells by FT-IR spectroscopy. *FEBS Lett.* **579**, 3433–3436.

Ami, D., Natalello, A., Taylor, G., Tonon, G., Doglia, S. M., (2006). Structural analysis of protein inclusion bodies by Fourier transform infrared microspectroscopy. *Biochim. Biophys. Acta Proteins Proteomics* **1764**, 793–799.

Anfinsen, C. B., (1973). Principles that govern folding of protein chains. *Science* **181**, 223–230.

Anson, M. L., Mirsky, A. E., (1931). The reversibility of protein coagulation. *J. Phys. Chem.* **35**, 185–193.

Arie, J. P., Miot, M., Sassoon, N., Betton, J. M., (2006). Formation of active inclusion bodies in the periplasm of *Escherichia coli*. *Mol. Microbiol.* **62**, 427–437.

Astbury, W. T., Dickinson, S., Bailey, K., (1935). The X-ray interpretation of denaturation and the structure of the seed globulins. *Biochim. J.* **29**, 2351–2360.

Baneyx, F., Mujacic, M., (2004). Recombinant protein folding and misfolding in *Escherichia coli*. *Nat. Biotechnol.* **22**, 1399–1408.

Barnhart, M. M., Chapman, M. R., (2006). Curli biogenesis and function. *Annu. Rev. Microbiol.* **60**, 131–147.

Baxa, U., Cassese, T., Kajava, A. V., Steven, A. C., (2006). Structure, Function, and Amyloidogenesis of Fungal Prions: Filament Polymorphism and Prion Variants. Fibrous Proteins: Amyloids, Prions and Beta Proteins. vol. 73. Elsevier Academic Press Inc, San Diego, p. 125.

Baxa, U., Taylor, K. L., Wall, J. S., Simon, M. N., Cheng, N. Q., Wickner, R. B., et al., (2003). Architecture of ure2p prion filaments—the N-terminal domains form a central core fiber. *J. Biol. Chem.* **278**, 43717–43727.

Bennett, M. J., Choe, S., Eisenberg, D., (1994). Domain swapping—entangling alliances between proteins. *Proc. Natl. Acad. Sci. U.S.A.* **91**, 3127–3131.

Bennett, M. J., Schlunegger, M. P., Eisenberg, D., (1995). 3D domain swapping—a mechanism for oligomer assembly. *Protein Sci.* **4**, 2455–2468.

Betton, J. M., Hofnung, M., (1996). Folding of a mutant maltose-binding protein of *Escherichia coli* which forms inclusion bodies. *J. Biol. Chem.* **271**, 8046–8052.

Betts, S., Haase-Pettingell, C., Cook, K., King, J., (2004). Buried hydrophobic side-chains essential for the folding of the parallel beta-helix domains of the P22 tailspike. *Protein Sci.* **13**, 2291–2303.

Betts, S., Haase-Pettingell, C., King, J., (1997). Mutational effects on inclusion body formation. In: Advances in Protein Chemistry. vol. 50. Frederic M. Richards, David S. Eisenberg and Peter S. Kim, (Eds.), Academic Press Inc., San Diego, pp. 243–264.

Betts, S., King, J., (1999). There's a right way and a wrong way: *in vivo* and *in vitro* folding, misfolding and subunit assembly of the P22 tailspike. *Struct. Fold Des.* **7**, R131–R139.

Betts, S., Speed, M., King, J., (1999). Detection of early aggregation intermediates by native gel electrophoresis and native Western blotting. *Amyloid, Prions, and Other Protein Aggregates, vol.* **309**, pp. 333–350.

Booth, D. R., Sunde, M., Bellotti, V., Robinson, C. V., Hutchinson, W. L., Fraser, P. E., et al., (1997). Instability, unfolding and aggregation of human lysozyme variants underlying amyloid fibrillogenesis. *Nature* **385**, 787–793.

Bousset, L., Melki, R., (2002). Similar and divergent features in mammalian and yeast prions. *Microbes Infect.* **4**, 461–469.

Bowden, G. A., Paredes, A. M., Georgiou, G., (1991). Structure and morphology of protein inclusion-bodies in *Escherichia coli*. *Bio-Technology* **9**, 725–730.

Bucciantini, M., Giannoni, E., Chiti, F., Baroni, F., Formigli, L., Zurdo, J. S., et al., (2002). Inherent toxicity of aggregates implies a common mechanism for protein misfolding diseases. *Nature* **416**, 507–511.

Butko, P., Buford, J. P., Goodwin, J. S., Stroud, P. A., McCormick, C. L., Cannon, G. C., (2001). Spectroscopic evidence for amyloid-like interfacial self-assembly of hydrophobin sc3. *Biochem. Biophys. Res. Commun.* **280**, 212–215.

Carrio, M., Gonzalez-Montalban, N., Vera, A., Villaverde, A., Ventura, S., (2005). Amyloid-like properties of bacterial inclusion bodies. *J. Mol. Biol.* **347**, 1025–1037.
Caughey, B., Lansbury, P. T., (2003). Protofibrils, pores, fibrils, and neurodegeneration: separating the responsible protein aggregates from the innocent bystanders. *Annu. Rev. Neurosci.* **26**, 267–298.
Chapman, M. R., Robinson, L. S., Pinkner, J. S., Roth, R., Heuser, J., Hammar, M., et al., (2002). Role of *Escherichia coli* curli operons in directing amyloid fiber formation. *Science* **295**, 851–855.
Chiti, F., Webster, P., Taddei, N., Clark, A., Stefani, M., Ramponi, G., et al., (1999). Designing conditions for *in vitro* formation of amyloid protofilaments and fibrils. *Proc. Natl. Acad. Sci. U.S.A.* **96**, 3590–3594.
Claessen, D., Rink, R., de Jong, W., Siebring, J., de Vreugd, P., Boersma, F. G.H., et al., (2003). A novel class of secreted hydrophobic proteins is involved in aerial hyphae formation in *Streptomyces coelicolor* by forming amyloid-like fibrils. *Genes Dev.* **17**, 1714–1726.
Colon, W., Kelly, J. W., (1992). Partial denaturation of transthyretin is sufficient for amyloid fibril formation in vitro. *Biochemistry* **31**, 8654–8660.
Colon, W., Lai, Z. H., McCutchen, S. L., Miroy, G. J., Strang, C., Kelly, J. W., et al., (1996). FAP mutations destabilize transthyretin facilitating conformational changes required for amyloid formation. In: Nature and Origin of Amyloid Fibrils. Ciba Foundation Symposia vol. 199. Bock G. R. and Goode J. A. (eds) John Wiley & Sons Ltd., Chichester, pp. 228–242.
Come, J. H., Fraser, P. E., Lansbury, P. T., (1993). A kinetic-model for amyloid formation in the prion diseases—importance of seeding. *Proc. Natl. Acad. Sci. U. S.A.* **90**, 5959–5963.
Connors, L. H., Shirahama, T., Skinner, M., Fenves, A., Cohen, A. S., (1985). In vitro formation of amyloid fibrils from intact beta-2-microglobulin. *Biochem. Biophys. Res. Commun.* **131**, 1063–1068.
Conway, K. A., Harper, J. D., Lansbury, P. T., (1998). Accelerated *in vitro* fibril formation by a mutant alpha-synuclein linked to early-onset parkinson disease. *Nat. Med.* **4**, 1318–1320.
Conway, K. A., Harper, J. D., Lansbury, P. T., (2000). Fibrils formed *in vitro* from alpha-synuclein and two mutant forms linked to Parkinson's disease are typical amyloid. *Biochemistry* **39**, 2552–2563.
Crestfield, A. M., Moore, S. W.H., (1963). Properties and conformation of the histidine residues at the active site of ribonuclease. *J. Biol. Chem.* **238**, 2421–2428.
Curtis-Fisk, J., Spencer, R. M., Weliky, D. P., (2008). Native conformation at specific residues in recombinant inclusion body protein in whole cells determined with solid-state NMR spectroscopy. *J. Am. Chem. Soc.* **130**, 12568.
De Bernardez Clark, E., Schwarz, E., Rudolph, R., (1999). Inhibition of aggregation side reactions during *in vitro* protein folding. *Methods Enzymol.* **309**, 217–236.
Diaz-Avalos, R., Long, C., Fontano, E., Balbirnie, M., Grothe, R., Eisenberg, D., et al., (2003). Cross-beta order and diversity in nanocrystals of an amyloid-forming peptide. *J. Mol. Biol.* **330**, 1165–1175.

Edskes, H. K., Gray, V. T., Wickner, R. B., (1999). The [URE3] prion is an aggregated form of ure2p that can be cured by overexpression of ure2p fragments. *Proc. Natl. Acad. Sci. U.S.A.* **96**, 1498–1503.

Fink, A. L., (1998). Protein aggregation: folding aggregates, inclusion bodies and amyloid. *Fold Des.* **3**, R9–R23.

Fowler, D. M., Koulov, A. V., Alory-Jost, C., Marks, M. S., Balch, W. E., Kelly, J. W., (2006). Functional amyloid formation within mammalian tissue. *PLos Biol.* **4**, 100– 107.

Fowler, D. M., Koulov, A. V., Balch, W. E., Kelly, J. W., (2007). Functional amyloid—from bacteria to humans. *Trends Biochem. Sci.* **32**, 217–224.

Fuchs, A., Seiderer, C., Seckler, R., (1991). Invitro folding pathway of phage-p22 tailspike protein. *Biochemistry* **30**, 6598–6604.

Fukuma, T., Mostaert, A. S., Jarvis, S. P., (2006). Explanation for the mechanical strength of amyloid fibrils. *Tribol. Lett.* **22**, 233–237.

Garcia-Fruitos, E., Gonzalez-Montalban, N., Morell, M., Vera, A., Ferraz, R. M., Aris, A., et al., (2005). Aggregation as bacterial inclusion bodies does not imply inactivation of enzymes and fluorescent proteins. *Microb. Cell Fact.* **4**, 27.

Garcia-Fruitos, E., Rodriguez-Carmona, E., Diez-Gil, C., Ferraz, R. M., Vazquez, E., Corchero, J. L., et al., (2009). Surface cell growth engineering assisted by a novel bacterial nanomaterial. *Adv. Mater.* **21**, 4249.

Gazit, E., (2007a). Self-assembled peptide nanostructures: the design of molecular building blocks and their technological utilization. *Chem. Soc. Rev.* **36**, 1263–1269.

Gazit, E., (2007b). Use of biomolecular templates for the fabrication of metal nanowires. *FEBS J.* **274**, 317–322.

Gebbink, M., Claessen, D., Bouma, B., Dijkhuizen, L., Wosten, H. A.B., (2005). Amyloids—A functional coat for microorganisms. *Nat. Rev. Microbiol.* **3**, 333–341.

Geddes, A. J., Atkins, P. K.D., Beighton, E. D.T., (1968). "Cross-beta" conformation in proteins. *J. Mol. Biol.* **32**, 343–358.

Gelain, F., Horii, A., Zhang, S. G., (2007). Designer self-assembling peptide scaffolds for 3-D tissue cell cultures and regenerative medicine. *Macromol. Biosci.* **7**, 544–551.

Georgiou, G., Valax, P., (1996). Expression of correctly folded proteins in *Escherichia coli*. *Curr. Opin. Biotechnol.* **7**, 190–197.

Gilead, S., Gazit, E., (2005). Self-Organization of Short Peptide Fragments: From Amyloid Fibrils to Nanoscale Supramolecular Assemblies. *Supramolecular Chemistry* **17**, 87–92.

Glenner, G. G., (1980a). Amyloid deposits and amyloidosis. The beta-fibrilloses (first of two parts). *N. Engl. J. Med.* **302**, 1283–1292.

Glenner, G. G., (1980b). Amyloid deposits and amyloidosis: the beta-fibrilloses (second of two parts). *N. Engl. J. Med.* **302**, 1333–1343.

Glenner, G. G., Eanes, E. D., Wiley, C. A., (1988). Amyloid fibrils formed from a segment of the pancreatic-islet amyloid protein. *Biochem. Biophys. Res. Commun.* **155**, 608–614.

Goldberg, M. E., Rudolph, R., Jaenicke, R., (1991). A kinetic-study of the competition between renaturation and aggregation during the refolding of denatured reduced egg-white lysozyme. *Biochemistry* **30**, 2790–2797.

Goldberg, M. E., Zetina, C. R. (1980). Importance of interdomain interactions in the structure, function and stability of the F1 and F2 domains isolated from the beta2 subumit of *E. coli* tryptophan synthase. *In*: "Protein Folding" (R. Jaenicke, Eds.), pp. 469–484. Elsevier-North Holland Biomedical Press. Amsterdam and New York.

Goldenberg, D., King, J., (1982). Trimeric intermediate in the *in vivo* folding and subunit assembly of the tail spike endorhamnosidase of bacteriophage P22. *Proc. Natl. Acad. Sci. U.S.A.* **79**, 3403–3407.

Goldenberg, D. P., Smith, D. H., King, J., (1983). Genetic analysis of the folding pathway for the tail spike protein of phage P22. *Proc. Natl. Acad. Sci. U.S.A.* **80**, 7060–7064.

Haase-Pettingell, C. A., King, J., (1988). Formation of aggregates from a thermolabile *in vivo* folding intermediate in P22 tailspike maturation. A model for inclusion body formation. *J. Biol. Chem.* **263**, 4977–4983.

Hammer, N. D., Schmidt, J. C., Chapman, M. R., (2007). The curli nucleator protein, CsgB, contains an amyloidogenic domain that directs CsgA polymerization. *Proc. Natl. Acad. Sci. U.S.A.* **104**, 12494–12499.

Hamodrakas, S. J., Paulson, J. R., Rodakis, G. C., Kafatos, F. C., (1983). X-ray-diffraction studies of a silkmoth chorion. *Int. J. Biol. Macromol.* **5**, 149–153.

Iconomidou, V. A., Chryssikos, G. D., Gionis, V., Vriend, G., Hoenger, A., Hamodrakas, S. J., (2001). Amyloid-like fibrils from an 18-residue peptide analogue of a part of the central domain of the B-family of silkmoth chorion proteins. *FEBS Lett.* **499**, 268–273.

Iconomidou, V. A., Hamodrakas, S. J., (2008). Natural protective amyloids. *Curr. Protein Pept. Sci.* **9**, 291–309.

Iconomidou, V. A., Vriend, G., Hamodrakas, S. J., (2000). Amyloids protect the silkmoth oocyte and embryo. *FEBS Lett.* **479**, 141–145.

Ivanova, M. I., Sawaya, M. R., Gingery, M., Attinger, A., Eisenberg, D., (2004). An amyloid-forming segment of beta 2-microglobulin suggests a molecular model for the fibril. *Proc. Natl. Acad. Sci. U.S.A.* **101**, 10584–10589.

Jaenicke, R., (1987). Folding and association of proteins. *Prog. Biophys. Mol. Biol.* **49**, 117–237.

Jaroniec, C. P., MacPhee, C. E., Bajaj, V. S., McMahon, M. T., Dobson, C. M., Griffin, R. G., (2004). High-resolution molecular structure of a peptide in an amyloid fibril determined by magic angle spinning NMR spectroscopy. *Proc. Natl. Acad. Sci. U.S.A.* **101**, 711–716.

Jarrett, J. T., Lansbury, P. T. Jr., (1993). Seeding "one-dimensional crystallization" of amyloid: a pathogenic mechanism in Alzheimer's disease and scrapie? *Cell.* **73**, 1055–1058.

Kane, J. F., Hartley, D. L., (1988). Formation of recombinant protein inclusion-bodies in *Escherichia coli*. *Trends Biotechnol.* **6**, 95–101.

Kasotakis, E., Mossou, E., Adler-Abramovich, L., Mitchell, E. P., Forsyth, V. T., Gazit, E., et al., (2009). Design of metal-binding sites onto self-assembled peptide fibrils. *Biopolymers* **92**, 164–172.

Kazantsev, A., Preisinger, E., Dranovsky, A., Goldgaber, D., Housman, D., (1999). Insoluble detergent-resistant aggregates form between pathological and nonpathological lengths of polyglutamine in mammalian cells. *Proc. Natl. Acad. Sci. U.S.A.* **96**, 11404–11409.

Kim, P. S., Baldwin, R. L., (1982). Specific intermediates in the folding recations of small proteins and the mechanism of protein folding pathways. *Annu. Rev. Biochem.* **51**, 459–489.

Kirschner, D. A., Abraham, C., Selkoe, D. J., (1986). X-ray-diffraction from intraneuronal paired helical filaments and extraneuronal amyloid fibers in Alzheimer-disease indicates cross-beta conformation. *Proc. Natl. Acad. Sci. U. S.A.* **83**, 503–507.

Lai, Z. H., Colon, W., Kelly, J. W., (1996). The acid-mediated denaturation pathway of transthyretin yields a conformational intermediate that can self-assemble into amyloid. *Biochemistry* **35**, 6470–6482.

Lansbury, P. T., (1997). Structural neurology: are seeds at the root of neuronal degeneration? *Neuron.* **19**, 1151–1154.

Lansbury, P. T. Jr., (1999). Evolution of amyloid: what normal protein folding may tell us about fibrillogenesis and disease. *Proc. Natl. Acad. Sci. U.S.A.* **96**, 3342–3344.

Larsen, P., Nielsen, J. L., Dueholm, M. S., Wetzel, R., Otzen, D., Nielsen, P. H., (2007). Amyloid adhesins are abundant in natural biofilms. *Environ. Microbiol.* **9**, 3077– 3090.

Linder, M. B., Szilvay, G. R., Nakari-Setala, T., Penttila, M. E., (2005). Hydrophobins: the protein-amphiphiles of filamentous fungi. *FEMS Microbiol. Rev.* **29**, 877–896.

Liu, Y., Eisenberg, D., (2002). 3D domain swapping: as domains continue to swap. *Protein Sci.* **11**, 1285–1299.

Liu, Y. S., Gotte, G., Libonati, M., Eisenberg, D., (2001). A domain-swapped RNase A dimer with implications for amyloid formation. *Nat. Struct. Biol.* **8**, 211–214.

London, J., Skrzynia, C., Goldberg, M. E., (1974). Renaturation of *Escherichia coli* tryptophanase after exposure to 8 M urea—evidence for existence of nucleation centers. *Eur. J. Biochem.* **47**, 409–415.

Luhrs, T., Ritter, C., Adrian, M., Riek-Loher, D., Bohrmann, B., Doeli, H., et al., (2005). 3D structure of Alzheimer's amyloid-beta (1–42) fibrils. *Proc. Natl. Acad. Sci. U.S.A.* **102**, 17342–17347.

Maji, S. K., Perrin, M. H., Sawaya, M. R., Jessberger, S., Vadodaria, K., Rissman, R. A., et al., (2009). Functional amyloids as natural storage of peptide hormones in pituitary secretory granules. *Science* **325**, 328–332.

Maji, S. K., Schubert, D., Rivier, C., Lee, S., Rivier, J. E., Riek, R., (2008). Amyloid as a depot for the formulation of long-acting drugs. *PLos Biol.* **6**, 240–252.

Makin, O. S., Atkins, E., Sikorski, P., Johansson, J., Serpell, L. C., (2005). Molecular basis for amyloid fibril formation and stability. *Proc. Natl. Acad. Sci. U.S.A.* **102**, 315–320.

Marston, F. A.O., (1986). The purification of eukaryotic polypeptides synthesized in *Escherichia coli. Biochem. J.* **240**, 1–12.

McCutchen, S. L., Colon, W., Kelly, J. W., (1993). Transthyretin mutation leu-55-pro significantly alters tetramer stability and increases amyloidogenicity. *Biochemistry* **32**, 12119–12127.

Mitraki, A., Danner, M., King, J., Seckler, R., (1993). Temperature-sensitive mutations and second-site suppressor substitutions affect folding of the P22 tailspike protein *in vitro*. *J. Biol. Chem.* **268**, 20071–20075.

Mitraki, A., Fane, B., Haase-Pettingell, C., Sturtevant, J., King, J., (1991). Global suppression of protein folding defects and inclusion body formation. *Science* **253**, 54–58.

Mitraki, A., King, J., (1989). Protein folding intermediates and inclusion body formation. *Bio/Technology* **7**, 690–699.

Morell, M., Bravo, R., Espargaro, A., Sisquella, X., Aviles, F. X., Fernandez-Busquets, X., et al., (2008). Inclusion bodies: specificity in their aggregation process and amyloid-like structure. *Biochim. Biophys. Acta Mol. Cell Res.* **1783**, 1815–1825.

Mostaert, A. S., Giordani, C., Crockett, R., Karsten, U., Schumann, R., Jarvis, S. P., (2009). Characterisation of amyloid nanostructures in the natural adhesive of unicellular subaerial algae. *J. Adhes.* **85**, 465–483.

Mostaert, A. S., Higgins, M. J., Fukuma, T., Rindi, F., Jarvis, S. P., (2006). Nanoscale mechanical characterisation of amyloid fibrils discovered in a natural adhesive. *J. Biol. Phys.* **32**, 393–401.

Nagarkar, R. P., Hule, R. A., Pochan, D. J., Schneider, J. P., (2010). Domain swapping in materials design. *Biopolymers* **94**, 141–155.

Nelson, R., Eisenberg, D., (2006). Recent atomic models of amyloid fibril structure. *Curr. Opin. Struct. Biol.* **16**, 260–265.

Nelson, R., Sawaya, M. R., Balbirnie, M., Madsen, A. O., Riekel, C., Grothe, R., et al., (2005). Structure of the cross-beta spine of amyloid-like fibrils. *Nature* **435**, 773–778.

Nilsson, M. R., (2004). Techniques to study amyloid fibril formation *in vitro*. *Methods* **34**, 151–160.

Oberg, K., Chrunyk, B. A., Wetzel, R., Fink, A. L., (1994). Native-like secondary structure in interleukin-1-beta inclusion-bodies by attenuated total reflectance FTIR. *Biochemistry.* **33**, 2628–2634.

Papanikolopoulou, K., Schoehn, G., Forge, V., Forsyth, V. T., Riekel, C., Hernandez, J. F., et al., (2005). Amyloid fibril formation from sequences of a natural beta- structured fibrous protein, the adenovirus fiber. *J. Biol. Chem.* **280**, 2481–2490.

Paravastu, A. K., Leapman, R. D., Yau, W. M., Tycko, R., (2008). Molecular structural basis for polymorphism in Alzheimer's beta-amyloid fibrils. *Proc. Natl. Acad. Sci. U.S.A.* **105**, 18349–18354.

Perutz, M. F. (1980). Unboiling an egg. In: R. Jaenicke, (Eds.), Protein Folding. Elsevier-North Holland Biomedical Press, p. 14. Amsterdam and New York.

Perutz, M. F., Johnson, T., Suzuki, M., Finch, J. T., (1994). Glutamine repeats as polar zippers—their possible role in inherited neurodegenerative diseases. *Proc. Natl. Acad. Sci. U.S.A.* **91**, 5355–5358.

Petkova, A. T., Ishii, Y., Balbach, J. J., Antzutkin, O. N., Leapman, R. D., Delaglio, F., et al., (2002). A structural model for Alzheimer's beta-amyloid fibrils based on experimental constraints from solid state NMR. *Proc. Natl. Acad. Sci. U.S.A.* **99**, 16742–16747.

Podrabsky, J. E., Carpenter, J. F., Hand, S. C., (2001). Survival of water stress in annual fish embryos: dehydration avoidance and egg envelope amyloid fibers. *Am. J. Physiol. Regul. Integr. Comp. Physiol.* **280**, R123–R131.

Prusiner, S. B., McKinley, M. P., Bowman, K. A., Bolton, D. C., Bendheim, P. E., Groth, D. F., et al., (1983). Scrapie prions aggregate to form amyloid-like birefringent rods. *Cell* **35**, 349–358.

Przybycien, T. M., Dunn, J. P., Valax, P., Georgiou, G., (1994). Secondary structure characterization of beta-lactamase inclusion-bodies. *Protein Eng.* **7**, 131–136.

Rajagopal, K., Schneider, J. P., (2004). Self-assembling peptides and proteins for nanotechnological applications. *Curr. Opin. Struct. Biol.* **14**, 480–486.

Rajan, R. S., Illing, M. E., Bence, N. F., Kopito, R. R., (2001). Specificity in intracellular protein aggregation and inclusion body formation. *Proc. Natl. Acad. Sci. U.S.A.* **98**, 13060–13065.

Reches, M., Gazit, E., (2003). Casting metal nanowires within discrete self-assembled peptide nanotubes. *Science* **300**, 625–627.

Rinas, U., Hoffmann, F., Betiku, E., Estape, D., Marten, S., (2007). Inclusion body anatomy and functioning of chaperone-mediated *in vivo* inclusion body disassembly during high-level recombinant protein production in *Escherichia coli*. *J. Biotechnol.* **127**, 244–257.

Rudolph, R., Zettlmeissl, G., Jaenicke, R., (1979). Reconstitution of lactic-dehydrogenase—non-covalent aggregation vs reactivation.2. Reactivation of irreversibly denatured aggregates. *Biochemistry* **18**, 5572–5575.

Ruzicka, Z., (1997). Protective role of the egg stalk in Chrysopidae (neuroptera). *Eur. J. Entomol.* **94**, 111–114.

Sambashivan, S., Liu, Y. S., Sawaya, M. R., Gingery, M., Eisenberg, D., (2005). Amyloid-like fibrils of ribonuclease A with three-dimensional domain-swapped and native-like structure. *Nature* **437**, 266–269.

Saul, F. A., Mourez, M., Vulliez-Le Normand, B., Sassoon, N., Bentley, G. A., Betton, J. M., (2003). Crystal structure of a defective folding protein. *Protein Sci.* **12**, 577– 585.

Saupe, S. J., (2007). A short history of small s: A prion of the fungus *Podospora anserina*. *Prion* **1**, 110–115.

Sawaya, M. R., Sambashivan, S., Nelson, R., Ivanova, M. I., Sievers, S. A., Apostol, M. I., et al., (2007). Atomic structures of amyloid cross-beta spines reveal varied steric zippers. *Nature* **447**, 453–457.

Scheibel, T., Parthasarathy, R., Sawicki, G., Lin, X. M., Jaeger, H., Lindquist, S. L., (2003). Conducting nanowires built by controlled self-assembly of amyloid fibers and selective metal deposition. *Proc. Natl. Acad. Sci. U.S.A.* **100**, 4527–4532.

Schein, C. H., (1989). Production of soluble recombinant proteins in bacteria. *Bio-Technology* **7**, 1141–1147.

Schuler, B., Rachel, R., Seckler, R., (1999). Formation of fibrous aggregates from a non-native intermediate: the isolated P22 tailspike beta-helix domain. *J. Biol. Chem.* **274**, 18589–18596.

Seckler, R., (1998). Folding and function of repetitive structure in the homotrimeric phage P22 tailspike protein. *J. Struct. Biol.* **122**, 216–222.

Serpell, L. C., (2000). Alzheimer's amyloid fibrils: structure and assembly. *Biochim. Biophys. Acta* **1502**, 16–30.

Serpell, L. C., Fraser, P. E., Sunde, M., (1999). X-ray fiber diffraction of amyloid fibrils. *Methods Enzymol.* **309**, 526–536.

Serpell, L. C., Smith, J. M., (2000). Direct visualisation of the beta-sheet structure of synthetic Alzheimer's amyloid. *J. Mol. Biol.* **299**, 225–231.

Shorter, J., Lindquist, S., (2005). Prions as adaptive conduits of memory and inheritance. *Nat. Rev. Genet.* **6**, 435–450.

Sinnott, S. B., Andrews, R., (2001). Carbon nanotubes: synthesis, properties, and applications. *Crit. Rev. Solid State Mater. Sci.* **26**, 145–249.

Sipe, J. D., (1992). Amyloidosis. *Annu. Rev. Biochem.* **61**, 947–975.

Sipe, J. D., Cohen, A. S., (2000). Review: history of the amyloid fibril. *J. Struct. Biol.* **130**, 88–98.

Smith, B. L., Schaffer, T. E., Viani, M., Thompson, J. B., Frederick, N. A., Kindt, J., et al., (1999). Molecular mechanistic origin of the toughness of natural adhesives, fibres and composites. *Nature* **399**, 761–763.

Speed, M. A., Morshead, T., Wang, D. I.C., King, J., (1997). Conformation of P22 tailspike folding and aggregation intermediates probed by monoclonal antibodies. *Protein Sci.* **6**, 99–108.

Speed, M. A., Wang, D. I. C., King, J., (1995). Multimeric intermediates in the pathway to the aggregated inclusion-body state for p22 tailspike polypeptide-chains. *Protein Sci.* **4**, 900–908.

Speed, M. A., Wang, D. I.C., King, J., (1996). Specific aggregation of partially folded polypeptide chains: the molecular basis of inclusion body composition. *Nat. Biotechnol.* **14**, 1283–1287.

Steinbacher, S., Seckler, R., Miller, S., Steipe, B., Huber, R., Reinemer, P., (1994). Crystal structure of P22 tailspike protein: interdigitated subunits in a thermostable trimer. *Science* **265**, 383–386.

Sunde, M., Blake, C., (1997). The structure of amyloid fibrils by electron microscopy and X-ray diffraction. *Adv. Protein Chem.* **50**, 123–159.

Sunde, M., Kwan, A. H.Y., Templeton, M. D., Beever, R. E., Mackay, J. P., (2008). Structural analysis of hydrophobins. *Micron* **39**, 773–784.

Sunde, M., Serpell, L. C., Bartlam, M., Fraser, P. E., Pepys, M. B., Blake, C. C., (1997). Common core structure of amyloid fibrils by synchrotron X-ray diffraction. *J. Mol. Biol.* **273**, 729–739.

Taylor, G., Hoare, M., Gray, D. R., Marston, F. A.O., (1986). Size and density of protein inclusion-bodies. *Bio-Technology* **4**, 553–557.

Teng, P. K., Eisenberg, D., (2009). Short protein segments can drive a non-fibrillizing protein into the amyloid state. *Protein Eng. Des. Sel.* **22**, 531–536.

Terrones, M., (2003). Science and technology of the twenty-first century: synthesis, properties and applications of carbon nanotubes. *Annu. Rev. Mater. Res.* **33**, 419–501.

Tokatlidis, K., Dhurjati, P., Beguin, P., (1993). Properties conferred on *Clostridium-thermocellum* endoglucanase celc by grafting the duplicated segment of endoglucanase celd. *Protein Eng.* **6**, 947–952.

Tokatlidis, K., Dhurjati, P., Millet, J., Beguin, P., Aubert, J. P., (1991). High-activity of inclusion-bodies formed in *Escherichia coli* overproducing *Clostridium-thermocellum* endoglucanase-D. *FEBS Lett.* **282**, 205–208.

Tokatlidis, K., Salamitou, S., Beguin, P., Dhurjati, P., Aubert, J. P., (1991). Interaction of the duplicated segment carried by *Clostridium-thermocellum* cellulases with cellulosome components. *FEBS Lett.* **291**, 185–188.

Ventura, S., Zurdo, J., Narayanan, S., Parreno, M., Mangues, R., Reif, B., et al., (2004). Short amino acid stretches can mediate amyloid formation in globular proteins: the src homology 3 (SH3) case. *Proc. Natl. Acad. Sci. U.S.A.* **101**, 7258–7263.

Waldo, G. S., Standish, B. M., Berendzen, J., Terwilliger, T. C., (1999). Rapid protein-folding assay using green fluorescent protein. *Nat. Biotechnol.* **17**, 691–695.

Wang, X., Chapman, M. R., (2008). Sequence determinants of bacterial amyloid formation. *J. Mol. Biol.* **380**, 570–580.

Wang, L., Maji, S. K., Sawaya, M. R., Eisenberg, D., Riek, R., (2008). Bacterial inclusion bodies contain amyloid-like structure. *PLos Biol.* **6**, 1791–1801.

Wang, X. A., Smith, D. R., Jones, J. W., Chapman, M. R., (2007). *In vitro* polymerization of a functional *Escherichia coli* amyloid protein. *J. Biol. Chem.* **282**, 3713–3719.

Wasmer, C., Benkemoun, L., Sabate, R., Steinmetz, M. O., Coulary-Salin, B., Wang, L., et al., (2009). Solid-state NMR spectroscopy reveals that E. coli inclusion bodies of HET-s(218–289) are amyloids. *Angew. Chem. Int. Ed.* **48**, 4858–4860.

Wetzel, R., (1994). Mutations and off-pathway aggregation of proteins. *Trends Biotechnol.* **12**, 193–198.

Wetzel, R., (1996). For protein misassembly, it's the "i" decade. *Cell* **86**, 699–702.

Wickner, R. B., Edskes, H. K., Shewmaker, F., Nakayashiki, T., Engel, A., McCann, L., et al., (2007). Yeast prions evolution of the prion concept. *Prion.* **1**, 94–100.

Wigley, W. C., Stidham, R. D., Smith, N. M., Hunt, J. F., Thomas, P. J., (2001). Protein solubility and folding monitored *in vivo* by structural complementation of a genetic marker protein. *Nat. Biotechnol.* **19**, 131–136.

Wurth, C., Guimard, N. K., Hecht, M. H., (2002). Mutations that reduce aggregation of the Alzheimer's abeta42 peptide: an unbiased search for the sequence determinants of abeta amyloidogenesis. *J. Mol. Biol.* **319**, 1279–1290.

Yemini, M., Reches, M., Gazit, E., Rishpon, J., (2005a). Peptide nanotube-modified electrodes for enzyme-biosensor applications. *Anal. Chem.* **77**, 5155–5159.

Yemini, M., Reches, M., Rishpon, J., Gazit, E., (2005b). Novel electrochemical biosensing platform using self-assembled peptide nanotubes. *Nano Lett.* **5**, 183–186.

Yu, M. H., King, J., (1984). Single amino-acid substitutions influencing the folding pathway of the phage-p22 tail spike endorhamnosidase. *Proc. Natl. Acad. Sci. U.S.A. Biol. Sci.* **81**, 6584–6588.

Yu, M. H., King, J., (1988). Surface amino-acids as sites of temperature-sensitive folding mutations in the p22 tailspike protein. *J. Biol. Chem.* **263**, 1424–1431.

Zettlmeissl, G., Rudolph, R., Jaenicke, R., (1979). Reconstitution of lactic-dehydrogenase—non-covalent aggregation vs reactivation.1. Physical-properties and kinetics of aggregation. *Biochemistry* **18**, 5567–5571.

Zettlmeissl, G., Rudolph, R., Jaenicke, R., (1981). Reconstitution of lactic-dehydrogenase after acid dissociation—the yield of reactivation is determined by conformational rearrangements of the dissociated monomers. *Eur. J. Biochem.* **121**, 169–175.

Zhang, S. G., (2003). Fabrication of novel biomaterials through molecular self-assembly. *Nat. Biotechnol.* **21**, 1171–1178.

TAKING CHARGE OF PROTEINS: FROM NEURODEGENERATION TO INDUSTRIAL BIOTECHNOLOGY

By BRYAN F. SHAW*, DEMETRI T. MOUSTAKAS,† JULIAN P. WHITELEGGE,** AND KYM F. FAULL††

*Department of Chemistry and Biochemistry, Baylor University, Waco, TX, USA
†Department of Computational Science, Infection Discovery, AstraZeneca Pharmaceuticals LP, Waltham, MA, USA
**Department of Psychiatry and Biobehavioral Sciences, UCLA, Los Angeles, CA, USA
††The Pasarow Mass Spectrometry Laboratory, NPI-Semel Institute, University of California, Los Angeles, CA, USA

Abstract

The aggregation and precipitation of a soluble protein—within a motor neuron, or a pharmaceutical vial, or even inside a industrial-scale hydrolysis chamber—is a problem in human health and in biotechnology.

DOI: 10.1016/S1876-1623(10)79004-0

A growing body of research is suggesting that the magnitude of the *net* charge of a protein is a determinant of the rate at which proteins self-assemble in solution into aggregates with amorphous or fibrillar (or uncharacterized) morphologies. This chapter discusses how this apparently *simple* electrostatic effect might explain—in part or entirely—the pathogenicity of some mutations that cause familial protein aggregation diseases—especially the familial forms of amyotrophic lateral sclerosis that are caused by mutations in the gene encoding superoxide dismutase-1 (SOD1). In parallel, this chapter *also* discusses how understanding these electrostatic effects can guide the engineering of *industrial* enzymes (such as α-amylase from *Bacillus licheniformis*) into forms that are more resistant to aggregation and thermal precipitation than the enzymes that are currently used, for example, in the production of ethanol from starch or cellulose.

I. One of the Great Challenges for Biochemists in the 21st Century: *Protein Aggregation*

The errant self-assembly of polypeptides into structures that are often rich in β-sheets (referred to in this chapter as *protein aggregation*) is now believed to be the fundamental cause of several human diseases, including type 2 diabetes, cataracts, cystic fibrosis, and many neurodegenerative diseases such as Alzheimer's and Parkinson's disease, and amyotrophic lateral sclerosis (ALS) (Chiti and Dobson, 2006). We still do not fully understand how the biophysical and physico-chemical properties of globular proteins and peptides—properties such as hydrophobicity, net charge, electrostatic surface potential and secondary and tertiary structure—facilitate the assembly of proteins into structures *that are toxic* (Chiti et al., 2003). A more complete understanding of the physical-organic chemistry of protein aggregation will better equip us to develop pharmaceutical therapies that specifically target misfolded proteins and modify their biophysical and physical-organic properties in ways that inhibit their aggregation.

Although preventing the inappropriate aggregation of proteins represents an important *biomedical* problem, protein aggregation also causes problems that are *not* directly relevant to human health and that exist outside the realm of biological systems. The establishment of sustainable sources of energy, for example—a problem that is economic, environmental, and even social in nature—is inhibited, in part, by protein aggregation. Up to 25% of the cost of producing fermentable sugars from lignocellulose stems from the cost of the cellulase enzyme (Lynd et al., 2008), which continually aggregates during the hydrolysis of the lignocellulose and which must therefore be frequently replenished during hydrolysis and fermentation. In spite of the fact that the annual global production of crop waste is nearly 74 teragrams (dry)—nearly half of which is in the form of rice straw that is often burned in open fields—there remains no cost-effective method for producing ethanol from lignocellulose (Torigoe et al., 2000; Sticklen, 2006). It is reasonable to expect, and the results of research discussed in this chapter are beginning to show, that much of what we learn about the fundamental biophysics of protein aggregation in the context of human disease—for example, *why* pathogenic variants aggregate faster than their wild-type counterparts—can be used to engineer enzymes that are more resistant to aggregation than their natural counterparts, and that are perhaps therefore more cost-effective for industry.

II. Almost All Proteins are Charged at Physiological pH

Proteins usually have a *net* electrostatic charge in solution, that is, they have more anionic than cationic functionalities, or vice versa. A protein will always have a net electrostatic charge, in aqueous media, when the pH does not equal the isoelectric point (pI) of the particular protein (Gitlin et al., 2006a): a protein will have a net negative charge at values of pH above its pI and a net positive charge at values of pH that are below the pI. Approximately one third of the amino acids at the protein–water interface of a typical globular protein are positively or negatively charged (Miller et al., 1987a, 1987b), and it has been estimated that 84% of proteins contained in the Protein Data Bank have a theoretical net charge (Z_{seq}) between -10 and $+10$ (Lawrence et al., 2007).

The theoretical net charge of a protein at physiological pH can be estimated from the sum of cationic residues (i.e., arginine and lysine) and anionic residues (i.e., aspartate and glutamate), while also accounting for the N- and C-termini and the possible acetylation of the N-terminus. Tightly bound organic or metallic ion cofactors should also be taken into account (Colton et al., 1997). The charge of histidine residues ($pK_a = 6.3$) is more difficult to predict in folded proteins but are typically considered neutral at physiological pH and not usually included when calculating the formal net charge of proteins at neutral pH. A survey of pK_a values of 39 different histidine residues from 13 different proteins (Edgcomb and Murphy, 2002), for example, showed that the pK_a of imidazole groups varied from the regularly cited value of 6.3 (as measured by NMR spectroscopy) (Thurlkill et al., 2006) and ranged from 4.6 to 9.2. This variation in pK_a is poorly understood, but loosely correlated with the distance of each histidine from solvent, and their electrostatic environment, as well as the degree to which they are involved in polar interactions with other functional groups (Edgcomb and Murphy, 2002).

Several types of posttranslational modifications occur in proteins (*in vivo*) that affect the net electrostatic charge of a protein, including the attachment of negatively charged polysaccharides (i.e., heparan sulfate); phosphorylation of serine, threonine, and tyrosine; deamidation of asparagine or glutamine; and acetylation of lysine or N-terminal. The intramolecular cross-linking of lysine and glutamine (catalyzed by transglutaminase enzymes) will also change the net charge of a protein, by neutralizing the positive charge of lysine. Certain types of oxidative damage to proteins, such as the carboxymethylation of lysine (an advanced glycation end product), will also change the net electrostatic charge of proteins.

III. The Net Charge of a Protein Affects Its Rate of Aggregation

The rate at which a protein aggregates can be a function of its net charge (either positive or negative). Perhaps the clearest demonstration of this correlation is the general observation that proteins aggregate more rapidly at pH values that are equal to their pI, where $Z_{seq} = 0$, than at pH values where they are charged (Pauling, 1993). Studies by several groups have shown that increasing the net charge of peptides or proteins (by site-directed mutagenesis or by the covalent modification of surface groups) can decrease the rate of aggregation of the protein *in vitro* relative to the "wild-type" or unmodified protein (Chiti et al., 2002, 2003; Lawrence et al., 2007; Shaw et al., 2008). It has also been suggested that some proteins are protected against aggregation *in vivo*, because they possess a high net charge, either positive or negative (Dyson and Wright, 2005; Sugase et al., 2007). *Natively disordered* proteins, for example, are proteins that are unfolded when expressed (and fold, *in vivo*, upon molecular recognition by a separate target protein). In spite of being unfolded, however, natively disordered proteins do not rapidly aggregate, as do many folded proteins when they populate unfolded states. Natively disordered proteins have been hypothesized to be resistant to aggregation because of their high net charge (Dyson and Wright, 2005; Sugase et al., 2007).

Protein engineers have only recently begun to exploit these apparently simple electrostatic effects by generating proteins with unnaturally high values of net charge (Vendruscolo and Dobson, 2007). Such proteins are remarkably resistant to aggregation (Lawrence et al., 2007; Shaw et al., 2008), and this resilience suggests that the net surface charge of a protein can affect—independently of the conformational stability or structure of the protein—the propensity of that protein to aggregate. Lawrence and Liu, for example, have generated "supercharged" variants of green fluorescent protein and glutathione-*S*-transferase by numerous rounds of site-directed mutagenesis; these unusual protein variants have changes in Z_{seq} between 20 and 50 units and are especially resilient to irreversible inactivation and aggregation —even after boiling (Lawrence et al., 2007). As discussed in more detail later in this chapter, by using covalent modifications instead of site-directed mutagenesis, Whitesides and coworkers have neutralized more than 17 lysine-ε-NH_3^+ groups in the hydrolytic enzyme α-amylase by acetylation and in doing so were able to increase the net negative charge of α-amylase by a factor of 10 (from −2 to −20) at pH 8.4. The highly charged chemical derivative of α-amylase was more resistant to irreversible inactivation and aggregation after exposure to denaturing conditions than the unmodified α-amylase enzyme (Shaw et al., 2008). In both of these examples, the secondary and tertiary structure (as measured by circular dichroism) and conformational stability (as measured by differential scanning calorimetry or unfolding in a chaotropic solution) of the highly charged proteins were not significantly different than the less charged counterparts; this structural and thermodynamic similarity suggests that the decreased rates of aggregation occurred as a result of the increased net charge of each protein.

IV. Misfolded or Mischarged?

A comparison of the biophysical and biochemical effects of pathogenic mutations on the gene products associated with several familial protein aggregation diseases has revealed that many mutations lower the conformational stability or alter the structure or the kinetics of folding of the expressed protein (Chiti and Dobson, 2006; Shaw and Valentine, 2007; Rumfeldt et al., 2009). Other mutations influence, instead, the posttranslational modification of the protein (e.g., by inhibiting the binding of a metal cofactor or increasing the rate of proteolysis of the protein, or changing the site of proteolysis). These types of perturbations to the structure, folding, conformational stability, and posttranslational modification of the expressed protein can account for the accelerated rate of aggregation of *most* protein variants that induce familial protein misfolding diseases. Nevertheless, a small fraction of the mutations that cause familial protein aggregation diseases have *no* apparent effect upon the structure, stability, folding kinetics, or posttranslational modification of the expressed protein. These inexplicable mutations do, however, decrease the formal net charge of the expressed protein (Chiti et al., 2002; Rodriguez et al., 2005; Shaw and Valentine, 2007).

One hypothesis that attempts to explain how these puzzling mutations promote the aggregation of the expressed protein is referred to in this chapter as the *electrostatic hypothesis.* This hypothesis postulates that these mutations increase the rate of protein aggregation by decreasing the net electrostatic charge of the expressed protein. It is also possible, even when a pathogenic mutation *does* alter the structure or lower the conformational stability of the protein, that a reduction in net charge may also contribute toward increasing the rate of aggregation of the protein.

V. The Case of Cu, Zn Superoxide Dismutase-1 and Amyotrophic Lateral Sclerosis

Mutations in the gene encoding superoxide dismutase-1 (SOD1) cause familial forms of the neurodegenerative disease amyotrophic lateral sclerosis (ALS; also known as Lou Gehrig's disease). This pathogenic set of mutations to SOD1 (Fig. 1) illustrates the electrostatic hypothesis by demonstrating the *need* for such a hypothesis (Rodriguez et al., 2005; Sandelin et al., 2007; Shaw and Valentine, 2007). The effects of ALS

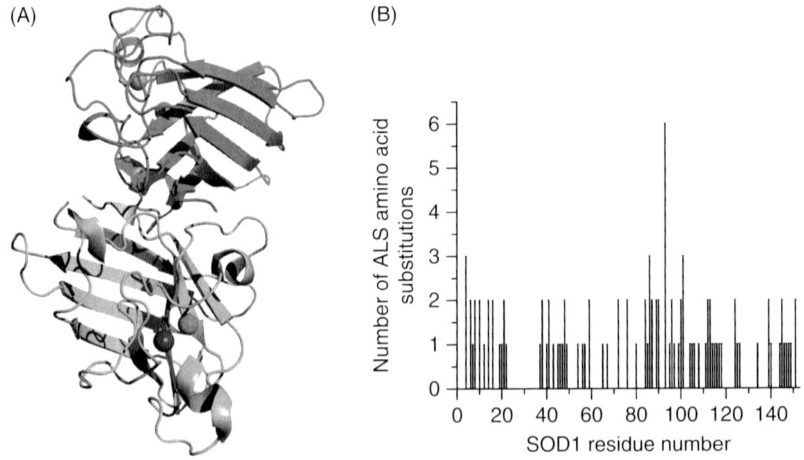

FIG. 1. (A) Ribbon diagram of human Cu, Zn superoxide dismutase 1 (SOD1). Different subunits of the homodimeric protein are colored orange and green. Each subunit binds one copper (cyan sphere) and one zinc ion (charcoal sphere). The fold of each subunit consists of an eight-stranded, antiparallel, Greek-key β-barrel (because the β-barrel of SOD1 is not completely closed, the fold of SOD1 is better described as a "β-taco"). PDB Code: 2V0A. (B) Most mutations in SOD1 that cause ALS (>90%) result in single amino acid substitutions. This frequency plot shows where each known ALS-linked amino acid substitution occurs in the 153-residue SOD1 polypeptide. The substitutions occur throughout the polypeptide chain, but glycine 93 appears to represent a "hot spot"; six different substitutions occur at Gly 93. (See color plate 7).

mutations in SOD1 upon the biochemical properties of the SOD1 protein are well understood: SOD1 is one of the best characterized proteins in biology (Hart et al., 1998; Assfalg et al., 2003; Banci et al., 2003, 2005; DiDonato et al., 2003; Elam et al., 2003; Shipp et al., 2003; Strange et al., 2003; Hough et al., 2004; Rodriguez et al., 2005; Potter et al., 2007; Roberts et al., 2007). A large amount of data—biochemical, biophysical, and structural—existed for SOD1 from as far back as the 1960s (McCord and Fridovich, 1969; Keele et al., 1971; Forman and Fridovich, 1973; Abernethy et al., 1974; Richardson et al., 1975) prior to the discovery of its involvement in ALS in 1993 (Rosen et al., 1993). The breadth of information for SOD1 exceeds that of many other proteins involved in other aggregation diseases, including for example, prion proteins, the htt protein, the amyloid precursor protein (APP), and α-synuclein. All of these proteins are causally linked to different protein aggregation diseases including the transmissible spongiform encephalopathies, amyloidosis, and Alzheimer's, Huntington's, and Parkinson's diseases, but the function, and in some cases the structures of these proteins, remains poorly understood (Harjes and Wanker, 2003; Vekrellis et al., 2004; Wang and Ha, 2004; Cattaneo et al., 2005; Zuccato et al., 2005; Koo, 2006; Millhauser, 2007).

The SOD1 protein is an antioxidant metalloenzyme that is ubiquitous in the cytosol, mitochondria, and nucleus of eukaryotic cells. SOD1 disproportionates two superoxide anions to molecular oxygen and hydrogen peroxide ($2O_2^- + 2H^+ \rightarrow H_2O_2 + O_2$). SOD1 is a homodimer that is comprised of two identical 16 kDa subunits (Fig. 1); each subunit coordinates one Cu and one Zn ion. The binding of Cu and Zn greatly increases the conformational stability of the protein: the coordination of 2Cu and 2Zn to homodimeric SOD1 increases the melting temperature from 52 °C to 95 °C at physiological pH (Valentine et al., 2005).

Amyotrophic lateral sclerosis is a fatal neurodegenerative disorder that is characterized by selective death of the motor neurons, followed by muscle atrophy and paralysis (Valentine et al., 2005). The progression of ALS is typically rapid: most patients die within 5 years of diagnosis, usually from asphyxiation. A quarter of the familial forms of ALS are caused by mutations in the gene encoding SOD1 (Rosen et al., 1993). Astonishingly, over 100 different mutations in SOD1 have now been identified to cause ALS. Most of these mutations (>90%) result in a single amino acid substitution to the 153 residue polypeptide; the others result in amino acid deletions, insertions, and C-terminal truncations. Intriguingly, the pathogenic amino acid substitutions are distributed throughout the amino acid sequence and are not clustered in any region or structural motif (Fig. 1). Several experiments have shown that ALS mutations to SOD1 induce pathogenesis by a "gain of function" mechanism, and this newly acquired function is considered to cause an increased propensity for the mutant SOD1 protein to aggregate. For example, transgenic mice lacking the endogenous mouse SOD1 (or human SOD1) do not develop ALS; transgenic mice expressing human WT SOD1 also do not develop ALS; transgenic mice expressing ALS-mutant forms of human SOD1 do, however, develop ALS and these mice also accumulate insoluble, aggregated forms of SOD1 in the tissues affected by ALS (Gurney et al., 1994; Wang et al., 2002).

Approximately 20 ALS-SOD1 variant proteins have been recombinantly expressed in *Saccharomyces cerevisiae, Escherichia coli*, or in insect cells and purified and characterized with one or more biophysical or structural tools including X-ray crystallography, NMR, differential scanning calorimetry, hydrogen–deuterium exchange, circular dichroism, analytical ultracentrifugation, electron paramagnetic resonance spectroscopy, and inductively coupled plasma atomic emission spectroscopy (Valentine et al., 2005). The biophysical characterization of ALS-variant SOD1 proteins has demonstrated that, like most pathogenic proteins involved in familial aggregation diseases, *most* ALS-variants have altered structures (Hart et al., 1998; Elam et al., 2003; Shipp et al., 2003; Strange et al., 2003; Hough et al., 2004; Antonyuk et al., 2005; Banci et al., 2005; Shaw et al., 2006; Roberts et al., 2007), kinetics of folding (Lindberg et al., 2004; Lindberg et al., 2005; Rumfeldt et al., 2009), decreased conformational stability (Hayward et al., 2002; Rodriguez et al., 2002; DiDonato et al., 2003; Rodriguez et al., 2005), or diminished affinity to bind Cu or Zn (Hayward et al., 2002; Rodriguez et al., 2002; Rodriguez et al., 2005; Valentine et al., 2005) compared to the nonpathogenic WT protein. Any one of these perturbations can result in a variant of SOD1 that has a faster rate of aggregation than the WT protein.

Out of the 20 ALS-variant proteins that have been studied in depth so far, there are 4 variants that do *not* differ significantly from the WT SOD1 in their 2° or 3° structure, thermostability, or the ability to coordinate metals. These variants are D90A, E100K, N139K, and D101N SOD1 (Fig. 2). All four of these amino acid substitutions reduce the formal net negative charge (Z_{seq}) of SOD1 at physiological pH (Fig. 2A) and in principle might influence the propensity of the protein to aggregate electrostatically (Shaw and Valentine, 2007). Decreases in Z_{seq} are also observed for other ALS mutations to SOD1 that have been shown experimentally to lower the conformational stability of SOD1 or diminish its affinity for Cu and Zn; this group of variants includes G85R, G37R, D125H, D124V, H46R, and E100G SOD1 (this list is likely to grow as more ALS variants of SOD1 are expressed and characterized). The decrease in Z_{seq} that is associated with each of these six amino acid substitutions will almost certainly contribute to increasing their rate of aggregation, in combination with the other effects of each substitution which also promotes aggregation (i.e., diminished thermostability or reduced affinity for metals).

All four of the cryptic substitutions D101N, D90A, E100K, and N139K, occur at the surface of SOD1 (Fig. 2). In order to determine the local surface electrostatic potential of each amino acid (e.g., D101, D90, E100,

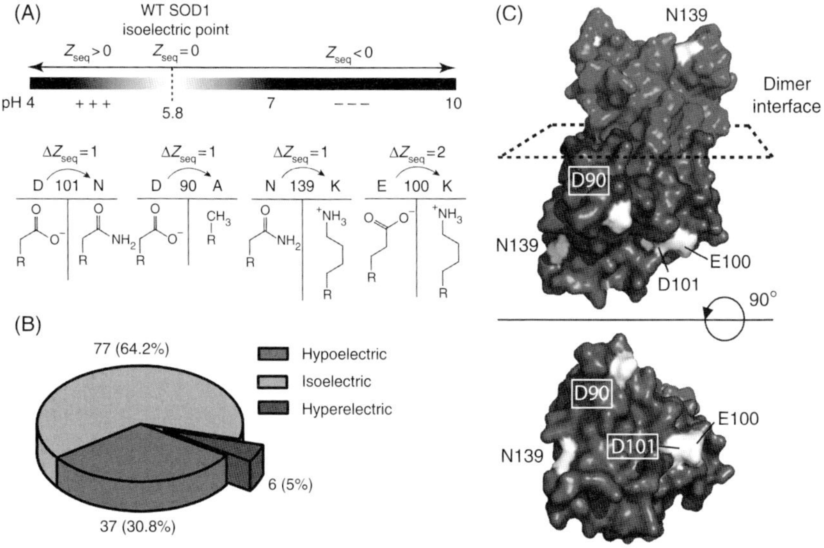

FIG. 2. (A) Wild-type (WT) human superoxide dismutase 1 (SOD1) is predicted to have a net negative charge at pH > 5.8. The calculated net charge (Z_{seq}) of the SOD1 monomer at pH 7.4 is −6 (for the homodimer $Z_{seq} = -12$). The *cryptic* mutants: D90A, D101N, N139K, and E100K do not significantly alter the structure, conformational stability, metal binding, or folding properties of the SOD1 protein. Each of these mutations does, however, decrease the formal net negative charge of SOD1 by 1 unit or 2 units at pH 7.4. (B) An analysis of 120 ALS mutations to SOD1 (available online at www.alsa.org) shows that 114 mutations (95%) are either isoelectric (e.g., do not change Z_{seq} at physiological pH) or hypoelectric (e.g., decrease Z_{seq}); only 5% are hyperelectric or increase Z_{seq} (1 of these 6 mutations, V7E, also severely destabilizes the folded SOD1 protein (Rodriguez et al., 2005)). (C) Surface rendering of the X-ray crystal structure of WT-SOD1 (PDB code: 2V0A). The D101N, D90A, N139K, and E100K substitutions occur at the surface of SOD1 and away from the dimer interface. (See color plate 8)

and N139) in wild-type SOD1, and to estimate how each of these four cryptic substitutions alters the electrostatic potential at the surface of SOD1, the nonlinear Poisson–Boltzmann equation was solved to calculate the electrostatic potential distribution at the solvent-accessible surface of each protein (Fig. 3). An X-ray crystal structure of the wild-type human SOD1 protein was used for these calculations (PDB code: 2V0A), and structures of each of the four mutants were modeled using the MOE software package (http://www.chemcomp.com); this modeling allowed the *in silico* introduction of amino acid substitutions in the crystal lattice and a

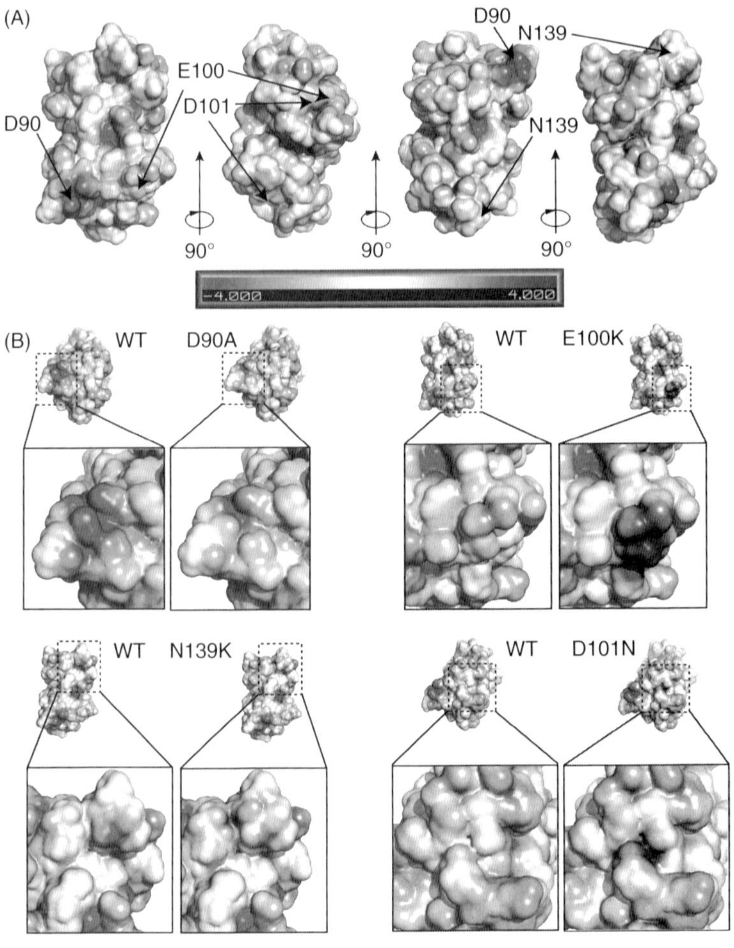

FIG. 3. (A) A nonlinear Poisson–Boltzmann equation was solved (using the APMS software package) to calculate the electrostatic potential distribution of the solvent accessible surface of wild-type human SOD1, based upon the X-ray crystal structure (PDB code: 2V0A); blue indicates positive electrostatic surface potential; red indicates negative electrostatic surface potential. (B) Structures of the 4 cryptic mutants (D90A, E100K, D101N, and N139K) were modeled by substituting each amino acid with wild-type SOD1 (e.g., the X-ray crystal structure of wild-type SOD1) using the MOE software package. We calculated the electrostatic potential distributions using the same method that we applied to the structure of wild-type SOD1. The unit of the color scale in both (A) and (B) is eV. These four cryptic substitutions do not alter the electrostatic potential of a similar region on the surface of SOD1 but affect different areas. (See color plate 9).

minimization of the energy of the resulting structure. The PyMOL software package (DeLano Scientific, Palo Alto, CA, USA) was then exploited to calculate the Poisson–Boltzmann distribution for each structure and to visualize the results (this version of PyMOL contains a graphical front-end to the APBS Poisson–Boltzmann solver (Baker et al., 2001)). No commonality emerged—either in the sign or magnitude of the electrostatic surface potential—at the points of substitution: D90 is surrounded by a large patch of uninterrupted negative potential on the surface of SOD1; E100 and D101 are located in a region of mixed positive and negative surface potential; and N139 is surrounded by a mix of neutral and negative surface potential (Fig. 3B). The results of the Poisson–Boltzmann calculations showed that these mutations tended to affect only the local electrostatic surface potential near the site of the substitution, that is, these pathogenic mutations do not uniformly affect the electrostatic potential of a similar motif or region of the SOD1 protein. The magnitude and polarity (e.g., sign) of the electrostatic surface potential of the area surrounding each substitution also appeared to be nonuniform: the D90A substitution reduced the magnitude of the negative surface potential surrounding residue 90, but the region continued to possess a local net negative surface potential after the substitution; the E100K and D101N substitutions increased the magnitude of positive surface potential of some proximal residues and also lowered the magnitude of negative surface potential of other proximal residues; the N139K substitution caused some residues with zero surface potential, to acquire a positive surface potential.

These calculations provide a picture of the electrostatic potential at the ground state of the pathogenic SOD1 proteins and suggest that the D90A, E100K, D101N, and N139K substitutions influence the electrostatic potential of different regions of the surface of the protein. Therefore, the only common biophysical effect of each mutation, upon the folded SOD1 protein, so far as we can presently see, is a reduction in the *net* negative charge of the entire protein at physiological pH.

VI. How Significant Are Reductions in the Net Charge of ALS-SOD1 Variants?

It must be remembered that the reductions in the net negative charge of SOD1 that are associated with many ALS-linked amino acid substitutions are *not* necessarily small effects and can result in a substantial decrease in the net charge of SOD1 (depending upon the solvent pH). The pI of SOD1 is approximately 5.8 and the formal net charge is estimated to be −6 at neutral pH. The net charge of human SOD1 has not yet been measured experimentally (surprisingly), but the net charge of SOD1 from *Bos taurus* was measured with capillary electrophoresis to be −4.8 at pH 8.4 ($Z_{seq} = -6.0$ for bovine SOD1 at pH 8.4) (Colton et al., 1997). The E100K mutation, wherein a negatively charged residue is replaced by a positively charged residue, will therefore reduce the formal net negative charge of SOD1 by 2 units—a reduction in Z_{seq} of 30% (Fig. 2). The other three cryptic mutations reduce the formal net negative charge of SOD1 by 1 unit (i.e., D90A: $- \rightarrow 0$; N139K: $0 \rightarrow +$; D101N: $- \rightarrow 0$).

VII. Cryptic Amino Acid Substitutions That Cause Other Familial Aggregation Diseases Reduce Net Charge

As mentioned earlier, there are examples of other familial protein aggregation diseases that are linked to a set of mutations, wherein a few mutations in the set accelerate the rate of aggregation of the expressed protein by an unknown mechanism. The E693K substitution in the APP, for example, causes cerebral hemorrhage and amyloidosis, but the mechanism by which this substitution promotes the aggregation of pathogenic fragment, Aβ is unknown (Chiti et al., 2003). The E693K substitution does *not* affect the posttranslational processing of the precursor protein (e.g., its proteolysis by γ-secretase) as do many pathogenic amino acid substitutions to APP, but the E693K substitution decreases the formal net charge of the pathogenic fragment, $A\beta_{1-40,\ 1-42}$, by 67% from −3 to −1 (Chiti et al., 2003). Another substitution in the same protein, and at the same location (E100G), reduces the net charge of the protein by only 1 formal unit and results in the development of early onset Alzheimer's disease instead of cerebral hemorrhage and amyloidosis. This substitution is similarly cryptic: the reason(s) why it increases the rate of aggregation of Aβ is unknown. A list of 16 other cryptic amino acid substitutions that cause 6 different hereditary protein aggregation diseases has been compiled by Dobson and coworkers who note that 14 of the substitutions result in decreases in the net charge of the expressed protein, or a decrease in the net charge of the proteolytic fragment that is thought to be a pathogenic species in each disease (Chiti et al., 2003).

VIII. Measuring the Net Charge of Proteins with Capillary Electrophoresis

When determining how an increase or decrease in the net charge of a protein might influence any of its properties, it is necessary to have a convenient experimental method for *measuring* the net charge of the protein. The actual net charge of a protein in solution can be quite different from the value that is predicted from the amino acid sequence (even when accounting for known posttranslational modifications). One explanation for this difference is an aberrant value for the pK_a of certain amino acids in the protein (compared to the standard pK_a for that amino acid). For example, glutamine 66 in Staphylococcal nuclease has a measured pK_a value of 8.8 (Garcia-Moreno et al., 1997). The standard value of pK_a for glutamate-γ-CO_2H is approximately 4.4 (Thurlkill et al., 2006) and Glu 66 will therefore *not* contribute a unit of negative charge to the net charge of nuclease at pH 7.4.

There are, unfortunately, few methods available for measuring the net charge of *folded* proteins under *physiological* conditions. One convenient method for measuring the net charge of a protein in solution is by measuring the electrophoretic mobilities of a "charge ladder" of the protein with capillary electrophoresis (Colton et al., 1997; Gao and Whitesides, 1997). A *protein charge ladder* is a set of derivatives of a protein that is generated by covalently modifying the functional groups at the surface with reagents that eliminate or introduce a charged functional group to the protein (Gitlin et al., 2006a). For most proteins, charge ladders are easily generated by acetylating the amino groups of lysine residues using acetic anhydride (Fig. 4A; an example of a charge ladder is shown in Fig. 6A). Each acetylation will neutralize the positive charge of lys-ε-NH_3^+ (e.g., lys-$NH_3^+ \rightarrow$ lys-$NHCOCH_3$). The change in the net charge that occurs from the acetylation of each lysine depends upon the pH but is not necessarily equal to -1.0. It has been determined, for example, that for bovine carbonic anhydrase II (BCA II)—a protein whose structure and activity is insensitive to the acetylation of its 18 lysine residues—each acetylation results in a change in net charge of -0.9 units. The difference from the integer value is due to the so-called "charge regulation"(Menon and Zydney, 2000) and is expected when considering that the lys-ε-NH_3^+ groups are not isolated functional groups but are a part of a network of ionizable functional groups (Gitlin et al., 2006a).

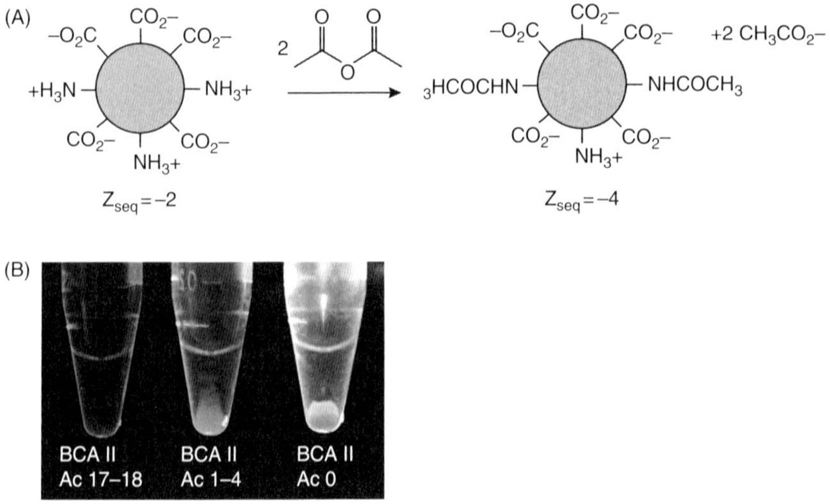

FIG. 4. (A) Neutralizing positive charges at the surface of a protein by acetylating lysine-ε-NH_3^+ with acetic anhydride can lead to large increases in the net negative charge of the protein. (B) Increasing the net charge of proteins by acetylating surface lysine residues can inhibit thermal aggregation. This image shows samples of bovine carbonic anhydrase II (BCA II) that have been acetylated to different degrees (e.g., Ac 0 = all lysine residues are unmodified; Ac 1–4: ≤4 lysines are acetylated; Ac 17–18: all 18 lysine residues are acetylated, with a minor fraction containing only 17 acetylated lysine residues). The peracetylation of lysine residues inhibited the formation of observable precipitate when the solution of protein was heated to 80 °C for 1 hour (followed by centrifugation at 16,000 × *g* for 2 minutes). The acetylation of between 1 and 4 lysine residues (i.e., Ac 1–4) results in a decrease in the amount of aggregated protein that is sedimentable at 16,000 × *g*. Because BCA II unfolds at 70 °C, and peracetylated BCA II unfolds at ~50 °C (Gitlin et al., 2006a), these particular experiments involved the thermally denatured proteins.

A bare, fused-silica capillary is used to separate the rungs of the protein charge ladder with capillary electrophoresis (CE); when a voltage is applied across the capillary, proteins separate according to their electrophoretic mobility (Grossman, 1992). Proteins with similar mass, shape, and the same net charge will band together, but those with similar mass and shape but different net charge will separate. The result is a ladder of bands with each band or 'rung' being comprised of proteins with an identical number of chemical modifications. The net charge of the unmodified protein can be calculated from the electrophoretic mobility of each rung according to the equation shown in Fig. 6A. In this way, the net charge of the *folded* protein can be directly measured under physiological conditions.

The ability to measure the net charge of a folded protein with capillary electrophoresis and charge ladders is in contrast to the capabilities of other techniques, for example, isoelectric focusing (IEF). The pI of an *unfolded* protein can be measured with IEF and the protein's net charge at any pH can be estimated from this value. The net charge of proteins, however, can be different in folded states compared to unfolded states: the values of pK_a of ionizable groups in a protein can in many cases be a function of the three-dimensional structure of the protein.

IX. Increasing the Durability of Enzymes with Organic Chemistry

During the course of experiments that used protein charge ladders to investigate how the surface chemistry of a protein affects its three dimensional structure, we discovered that the acetylation of all 18 lysine residues in BCA II completely protected the protein from thermal precipitation (Fig. 4B). We hypothesized that the acetylation of lysine residues could be used to generate highly charged—and catalytically active—derivatives of enzymes that are used in industrial biotechnology; we hypothesized that these highly charged derivatives would be more resistant to aggregation than the enzymes that are currently used for industrial purposes.

We chose to focus first on attempting to generate highly charged derivatives of the hydrolytic enzyme α-amylase. This well characterized and commercially available enzyme hydrolyzes α-1,4 glycosidic bonds in starch and is used in a surprisingly broad set of industrial applications (Kirk et al., 2002). For several of these industrial processes the enzyme must be active at values of pH above 6.0. The pI of the particular type of α-amylase that was used (isolated from the soil bacterium *Bacillus licheniformis*) is at pH 6.0, and thus the protein is expected to be negatively charged at alkaline pH, and each acetylation will *increase* the net negative charge of the enzyme. Also, because 28 of the 483 residues of *B. licheniformis* α-amylase (denoted hereafter as BLA) are lysine, this protein has the potential of having the net negative charge increased substantially via lysine acetylation. Several X-ray crystal structures exist for BLA and for homologous proteins from other organisms. Most of the 28 lysine residues in BLA (with the exception of Lys 234) are distal to the active site, and it was hypothesized that lysine acetylation would not adversely affect the enzymatic activity of BLA (Fig. 5).

We used a commercially available form of BLA that was previously engineered, with site-directed mutagenesis, to have an increased thermostability. This variant is identical to the wild-type protein with the exception of three amino acid substitutions that increase the surface hydrophobicity: N190F, Q264S, and N265Y. This engineered variant of BLA is referred to in this chapter as "BLA" without necessarily denoting these amino acid substitutions that increase the thermostability of BLA.

A protein charge ladder was prepared by adding different stoichiometric equivalents of acetic anhydride (between 0 and 200) to different aliquots of BLA, followed by the combination of each aliquot to yield the full ladder (Fig. 6A). Thus each rung of the ladder (e.g., each peak in Fig. 6A) represents a mixture of BLA polypeptides that have the same number of acetylated lysine residues, but each rung is likely to contain a heterogeneous mixture of regioisomers because lysine acetylation likely occurs in a nonselective manner. We were able to acetylate most of the lysine residues in BLA by using acetic anhydride. Twenty-three of the twenty-eight BLA

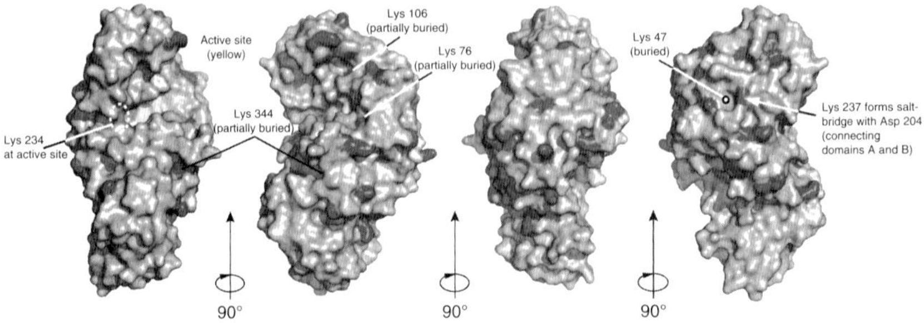

FIG. 5. Surface rendering of the X-ray crystal structure of the thermostable variant of α-amylase from *Bacillus licheniformis* (N190F/Q264S/N265Y BLA). Lysine residues are colored in red (with the exception of the strictly conserved lysine 234, at the active site, which is colored green). The active site region is colored yellow. The majority of lysine residues are not located on the same face of the protein as the active site, and nearly all 28 lysine residues are solvent exposed. Lysine 76, 106, and 344 are partially buried and Lys 47 is completely buried and not visible with a surface rendering. PDB code:1BLI. (See color plate 10).

lysine residues were readily modified by a stoichiometric excess of acetic anhydride, as demonstrated by the analysis of the charge ladder (Fig. 6A). The inability to acylate *all* 28 lysine residues—even in the presence of >100 equivalents of acetic anhydride—is not surprising given that the reaction was carried out in conditions that strongly favor the folded form of the enzyme (e.g., pH 9.0, 25 °C; the melting point of the unmodified protein is ~83 °C). The burial of some lysine residues from solvent or an intramolecular interaction involving lysine side chains could hinder the reactivity of the lysine-ε-NH_3^+ group with acetic anhydride. There are three lysine residues (Lys 76, 106, and 344) in BLA that appear from the X-ray crystal structure to be partially buried from solvent, but only a single lysine

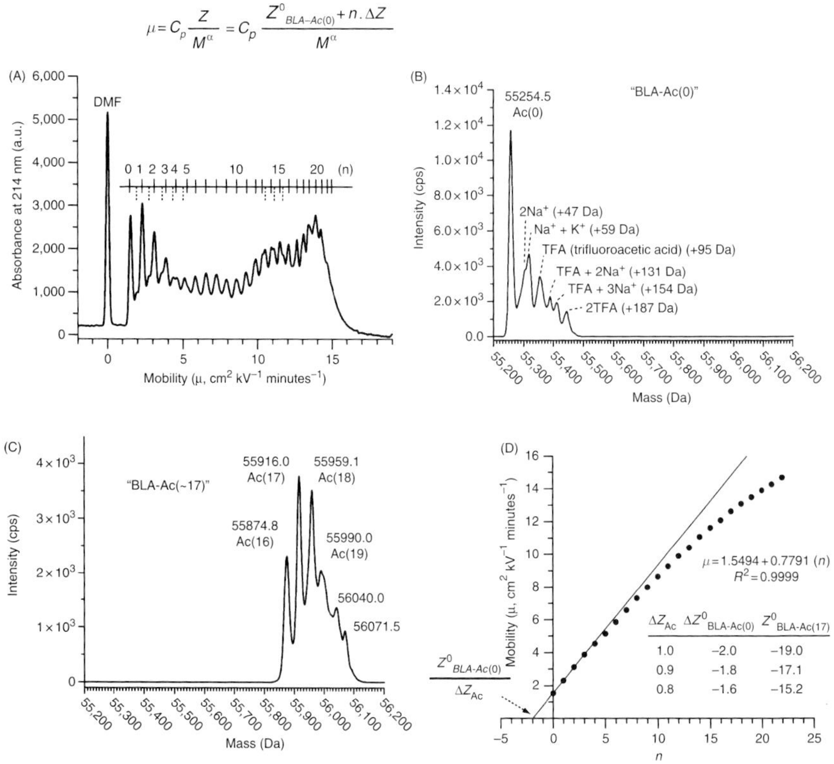

FIG. 6. (Continued)

residue (Lys 47) is completely buried (Fig. 5; structural features of all 28 lysine residues are summarized in Table I).

A particular batch of acetylated BLA that had between 16 and 21 acetylated lysine residues was targeted for detailed characterization; this sample was prepared by adding a specific stoichiometric amount of acetic anhydride to BLA. The resulting mixture was not purified further (e.g., a BLA protein with a specific number of acetylations was not isolated from the mixture). Figure 6B and 6C shows the electrospray ionization mass spectra of the unmodified BLA and this mixture of acetylated BLA. The predominant derivative in the mixture contains 17 acetylations; we therefore refer to this mixture as BLA-Ac(~17).

With the equation shown in Fig. 6A it was possible to calculate the net charge of unmodified BLA (denoted BLA-Ac(0)) from a plot of the values

---

FIG. 6. The net negative charge of a hydrolytic enzyme (α-amylase) can be increased by an order of magnitude by the acetylation of its lysine residues with acetic anhydride. (A) A capillary electropherogram of a charge ladder of a heat-stable variant of α-amylase from *Bacillus licheniformis* (denoted BLA) prepared by acetylating lysine-ε-NH_3^+ with acetic anhydride. This commercially available variant of BLA contains three amino acid substitutions (N190F, Q264S, and N265Y) that increase the thermostability of the protein. The charge ladder was prepared by successively acetylating lysine-ε-NH_3^+ (denoted Ac(n)) at pH 9, 23 °C. The "ruler" above the charge ladder correlates the number of acetyl modifications, n, with each rung of the ladder; dashed lines denote shoulders of rungs that constitute a faint second and third charge ladder (the nature of these ladders are unknown). Only between 20 and 23 lysine residues could be acetylated under native conditions (out of 28 lysine residues and 1 nonacetylated N-terminus). Dimethylformamide (DMF) is used as an electrically neutral marker. The electrophoretic mobility (μ) can be expressed as a function of net charge (Z); M = molecular weight; C_p and α are constant values; theoretical values of α vary from 0.3 to 1.0 and depend on the model used to derive the hydrodynamic drag from the molecular weight; typically for globular proteins $\alpha \approx 2/3$ (Rickard et al., 1991; Basak and Ladisch, 1995). (B and C) Electrospray ionization mass spectra of nonacetylated N190F/Q264S/N265Y BLA (denoted "BLA-Ac(0)") and acetylated derivatives (denoted "BLA-Ac(~17)"). This solution of acetylated protein is a mixture of variably acetylated species but consists of two predominant species: BLA-Ac(17) and BLA-Ac(18). (D) The electrophoretic mobility (μ) of each rung of the charge ladder of BLA plotted as a function of n (e.g., the number of acetyl modifications). A line was fit to the linear region of the plot; the x-intercept is equal to the ratio between the net charge of the rung of lowest mobility ($Z^\circ_{BLA\text{-}Ac(0)}$) and the change in charge resulting from a single acetylation (ΔZ_{Ac}). The net charge (Z^o) of BLA-Ac(0) ($\mu = 1.3\,cm^2\,kV^{-1}$ minutes^{-1}) is therefore approximately equal to the x-intercept. The inset table shows different possible values of $Z^\circ_{BLA\text{-}Ac(0)}$ as a function of ΔZ_{Ac} (the value of this latter term typically varies from 0.8 to 1.0). Adapted from Shaw and Whitesides (Shaw et al., 2008).

TABLE I
The Susceptibility of Lysine Residues (Lys) to Acetylation (Ac) in α-amylase from *Bacillus licheniformis*: A Comparison of Data from X-ray Crystallography and Bottom-Up Proteomics

Lys	Ac	Buried	Ionic interaction[a]	Lys	Ac	Buried	Ionic interaction
α23	N/A[b]	No	Glu82	α237	Yes	No	Asp204
Π47	No	Yes	Asp63	α251	Yes	No	Glu250
Π70	Yes	No	None	Π254	?	No	None
Π76	No	Partially	None	α276	?	No	Glu185
α80	Yes	No	Glu66, Glu222	α306	Yes	No	None
α88	Yes	No	None	α315	Yes	No	Glu271
Π106	No	Partially	None	Π319	Yes	No	None
β136	N/A	No	Asp114	α344	No	Partially	Asp 325, Asp365
Π154	Yes	No	None	Π370	Yes	No	None
Π170	N/A	No	Glu167	α381	Yes	No	Glu385
β176	Yes	No	None	α383	Yes	No	None
Π180	No	No	Asp 202, Asp 121	α389	Yes	No	None
α213	Yes	No	None	α392	Yes	No	None
α234[c]	Yes	No	Asp204	β436	Yes	No	Asp404

The symbols α, β, and Π are used to indicate whether the specific lysine residue is contained within an α-helix, β-strand, or in a loop (Π) in the crystal structure of BLA (PDB code 1BLI).

[a] Ionic interactions reported in three studies of metalated (e.g., Ca_3Na) and apo (e.g., Ca, Na-free) BLA with X-ray crystallography.

[b] N/A denotes that the specific lysine residue was not included in any of the proteolytic peptides identified by tandem mass spectrometry and that the acetylation status could not be assigned.

[c] Lysine 234 is conserved in α-amylase proteins from *B. licheniformis, Bacillus stearothermophilus*, ovine pancreas, and *Aspergillus oryzae*; the 27 other lysine residues in BLA are not conserved. "?" denotes ambiguous assignment in the mass spectrometric data set. X-ray crystallographic (Machius et al., 1995; Hwang et al., 1997; Machius et al., 1998) and mass spectrometric data (Shaw et al., 2008) are from published sources.

of mobility of each rung of the BLA charge ladder as a function of the number of acetyl modifications (denoted "*n*" in Fig. 6D). The net charge of BLA-Ac(0) was calculated to be between −1.6 and −2.0 at pH 8.4; the range in the value of net charge is due to the uncertainty of the value ΔZ that expresses the change in charge that results from each modification (see inset Table in Fig. 6D). The net charge of BLA-Ac(~17) was calculated to be between −15.2 and −19.0, again depending upon the range in the value of ΔZ (from −0.8 to −1.0). Therefore, the acetylation of BLA with a stoichiometric excess of acetic anhydride increased the net negative charge of BLA by nearly an order of magnitude.

X. The Effects of Acetylation on the Activity, Secondary Structure, and Thermostability of BLA

The structure of BLA-Ac(~17) appears to be similar to that of BLA-Ac(0), in spite of the fact that 60% of the lysine residues are acetylated. Figure 7A compares the superimposable circular dichroic spectra for the acetylated and unmodified proteins. BLA is not unique in its ability to be exhaustively acetylated without a change in structure. As already mentioned, BCA II—a thermostable metalloprotein—can be *peracetylated* (e.g., all 18 lysine residues can be acetylated) without a significant change in the secondary and tertiary structure of the folded protein. The peracetylated derivative of BCA II has been characterized in greater detail than the acylated derivatives of BLA, and BCA II is one example that illustrates how the acetylation of lysine can be used to change the surface charge of an enzyme, without affecting the structure and activity of the enzyme (Gitlin et al., 2006a). The peracetylated form of BCA II has, for example, recently been crystallized and analyzed by X-ray crystallography. The backbone of the peracetylated and unmodified proteins are nearly superimposable (Snyder, unpublished observation); the root mean square deviation (RMSD) of all atoms in the crystal structure of peracetylated and unmodified BCA II is 0.294 Å; CD spectra for peracetylated BCA II and the unmodified protein are also similar (Gitlin et al., 2006a; Krishnamurthy et al., 2008).

The enzymatic activity of BLA is also not affected by the acetylation of most of its lysine residues. We measured the enzymatic activity of BLA-Ac (~17) by measuring the rate of hydrolysis of a solution of 4,6-*o*-benzilidine-*p*-nitrophenyl maltoheptaoside in the presence of glucosidase (e.g., the so-called Ceralpha assay (McCleary et al., 2002)); the rate of hydrolysis can be measured by an increase in absorbance at 400 nm. We performed our enzymatic assays by diluting aliquots of BLA-Ac(0) and BLA-Ac(~17) from solutions of surfactant at concentrations that denature BLA (e.g., 100-mM SDS) into solutions containing no surfactant and, in this way, we also examine the ability of the highly acetylated protein to refold properly. Figure 7A shows that both BLA-Ac(~17) and BLA-Ac(0) hydrolyze the oligosaccharide at a similar rate after refolding from high concentrations (100 mM) of SDS.

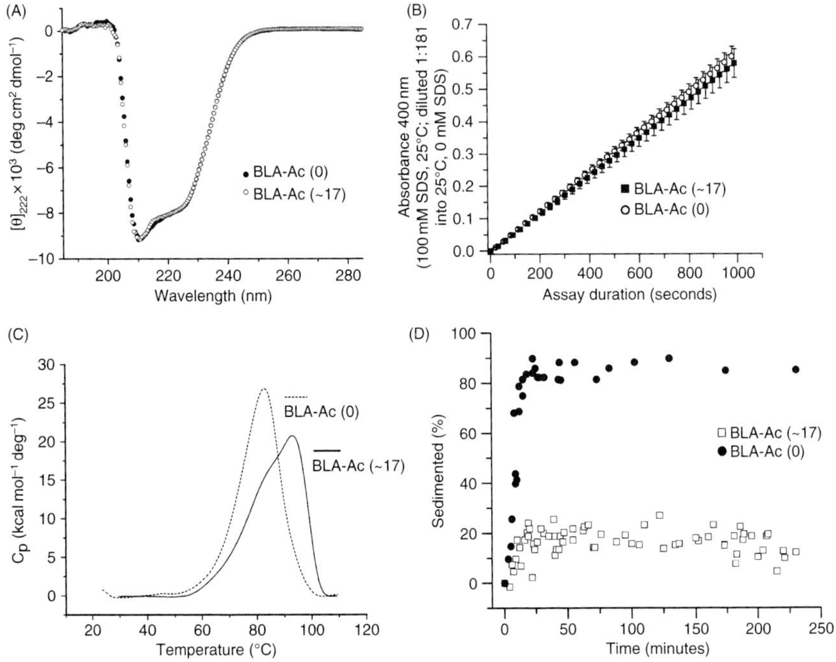

FIG. 7. Acetylating more than a dozen lysine residues in N190F/Q264S/N265Y BLA does not decrease the enzymatic activity or thermostability nor perturb the structure of the protein. Acetylation does, however, inhibit the thermal aggregation of BLA. (A) Analysis of unmodified BLA (denoted "BLA-Ac(0)") and acetylated BLA (denoted "BLA-Ac(~17)") with circular dichroism. CD spectra of acetylated and nonacetylated proteins were collected at pH 8.5, 25-mM Tris, 25 °C. (B) Enzymatic activities of BLA-Ac (~17) and BLA-Ac(0) were equal after exposure to 100-mM SDS at room temperature (pH 8.5, 25-mM Tris); the enzymatic activity was determined after proteins were diluted 1:181 from SDS solutions into Ceralpha assay solutions containing no surfactant. Error bars represent the standard deviation of absorbance values averaged from the mean ($N = 7$). Activity was measured using an α-gluocosidase/4,6-*o*-benzilidine-*p*-nitrophenyl maltoheptaoside procedure (Shaw et al., 2008). (C) Thermal denaturation of BLA-Ac (~17) and BLA-Ac(0) monitored by differential scanning calorimetry at pH 7.4, 10-mM phosphate. The thermogram for BLA-Ac(0) had a single transition at $T_m = 83$ °C; the thermal denaturation of BLA-Ac(~17) yielded two endothermic transitions at 84 and 93 °C. (D) Lysine acetylation inhibited BLA from forming sedimentable aggregates under denaturing conditions. Solutions of BLA-Ac(0) and BLA-Ac(~17) (9-μM BLA, pH 8.5, 10% trifluoroethanol) were heated to 90 °C and aliquots were removed at various times and centrifuged at 16,000 × *g*. Supernatant solutions were measured with UV-Vis spectroscopy (280 nm) to determine the amount of BLA protein and remained in the supernatant and did not aggregate into a sedimentable species. The data for BLA-Ac(0) and BLA-Ac(~17) are combined from three separate experiments each. Adapted from Shaw and Whitesides (Shaw et al., 2008).

We anticipated, based upon the results of similar experiments with other proteins (Jain and Hamilton, 2002), that the acetylation of so many lysine residues in BLA would result in a significant decrease in the thermostability of BLA. We were, however, wrong. An analysis of BLA-Ac(~17) and BLA-Ac(0) with differential scanning calorimetry revealed that acetylation actually *increased* (slightly) the thermostability of BLA. The temperature of unfolding of BLA-Ac(0) was measured to be 83 °C, and the thermogram was characterized by a single transition. The thermal unfolding of BLA-Ac(~17) yielded, in contrast, an endotherm with two partially resolved transitions: one at 84 °C and the other at 93 °C (Fig. 7C). The significance of the two transitions remains enigmatic. Remembering that BLA-Ac(~17) is a mixture likely comprised of different regioisomers, it is speculated that two transitions are observed for BLA-Ac(~17) because of a decrease in the cooperativity of unfolding of some rungs, and not others, or because certain regioisomers might have distinctively different temperatures of unfolding. Both of these scenarios could contribute to the endothermic transition shown in Fig. 7C.

The slightly higher temperature of unfolding for BLA-Ac(~17), than BLA-Ac(0), is surprising because it appears to violate a general rule of thumb in protein engineering: amino acid substitutions that increase the surface hydrophobicity of a protein will, generally, decrease the conformational stability of the protein (Matthews, 1993; Lee and Vasmatzis, 1997). Although the acetylation of lysine-ε-NH_3^+ increases the net negative charge, this reaction also increases the surface hydrophobicity. The Hansch Π-parameter (log P) for an NH_3^+ group is log $P = -2.12$ and log $P = -1.21$ for $NHCOCH_3$ groups (Hansch and Steward, 1964; Hansch and Coats, 1970) Nevertheless, one protein that represents a well-known exception to this rule of thumb in protein engineering is α-amylase from *B. licheniformis* (Declerck et al., 2003; Machius et al., 2003). More than nine different hydrophobic variants of BLA have been engineered with site directed mutagenesis, wherein each amino acid substitution involves a surface residue and an increase in the surface hydrophobicity. The hepta-variant H133I/H156Y/A181T/N190F/A209V/Q264S/N265Y has, for example, a $T_m = 106$ °C—a value that is more than 20° higher than that of the wild-type protein. It has been hypothesized that the substitution of surface residues with residues that are more bulky (e.g., more hydrophobic) results in an optimization of the molecular packing at indentations that exist at the surface of BLA (Declerck et al., 2003). We conclude, therefore, that the acetylation of the numerous lysine residues in BLA results in an increase in the thermostability of the enzyme because the reactions increase the hydrophobicity of the surface (Declerck et al., 2003). Likewise, it is concluded that acetylation of BLA lysine residues does *not* lead to any increase in the thermostability of the enzyme because of any type of *electrostatic* effect.

XI. The Highly Charged BLA Protein Resists Thermal Precipitation

Solutions containing unmodified BLA quickly form a precipitate when heated near the boiling point of water. Figure 7D shows that the heating of a solution of BLA-Ac(0) to approximately 90 °C in the presence of an organic cosolvent (trifluoroethanol) resulted in the rapid precipitation of BLA-Ac(0). Protein aggregation was measured by removing aliquots of solution as a function of time and measuring the protein concentration that remained in solution after centrifugation (16,000 × 2; 2 minutes). The BLA protein that remains soluble is presumably not aggregated. Figure 7D shows that most of the BLA-Ac(0) protein (~90%) no longer remained soluble after 25 minutes. The highly charged BLA-Ac(~17) protein is, however, much more resistant to thermal- and chemical-mediated precipitation. A fraction of the solution of BLA-Ac(~17)—approximately 10–20%—underwent aggregation after only 25 minutes, but the remainder was resilient to aggregation for extended periods of time. Even after 4 hours at 90 °C, approximately 80% of the BLA-Ac(~17) protein remained soluble and, presumably, not aggregated. The decreased rate of aggregation of BLA-Ac(~17), compared to BLA-Ac(0) is attributed to the increased net negative charge of the acetylated protein.

XII. Industrial Applications for Highly Charged Enzymes

Although the use of enzymes for generating combustible fuels from biomass is one of the most important applications of enzymes in biotechnology, the reader might be surprised to learn that currently the largest market for purified enzymes (by both volume and cost) is *laundry detergent*. The process of cleaning our clothes—although it might not seem as pressing of a matter as the development and production of sustainable sources of energy —consumes a surprisingly large amount of energy. Concerted efforts are being made to develop enzymatic detergents that are more environmentally friendly and that work efficiently at low temperatures, obviating the requirement of hot water (and enzyme-surfactant solutions that can be packaged and transported in more highly concentrated forms). Enzymatic detergents have the potential to fulfill these requirements, provided the enzymes do not become irreversibly inactivated when stored in high concentrations of detergent. To these ends, we have tested the enzymatic activity of BLA-Ac(~17) in different solutions of detergent and have compared its half-life to that of unmodified BLA (Fig. 8).

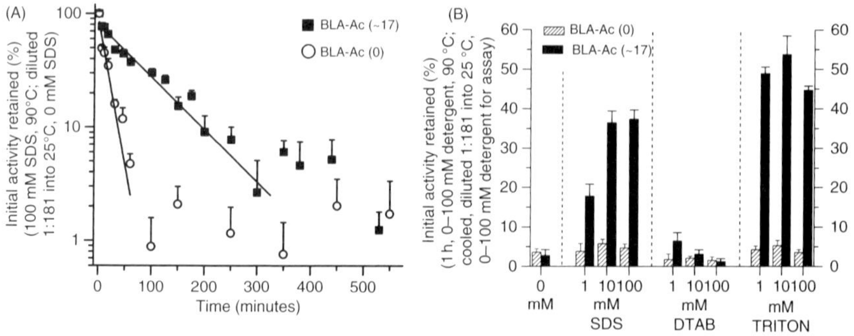

FIG. 8. The irreversible inactivation of BLA under denaturing conditions (90 °C, 100-mM SDS, 25-mM Tris, 192-mM glycine, pH 8.5) is inhibited by acetylation of its lysine residues. Activity assays were initiated by diluting an aliquot of heated BLA solutions (100-mM SDS, 90 °C) into the α-gluocosidase/4,6-*o*-benzilidine-*p*-nitrophenyl maltoheptaoside Ceralpha assay solution (1:181 v/v). (A) Inactivation of BLA-Ac(~17) and BLA-Ac(0) after incubation in 100-mM SDS (90 °C, pH 8.5), expressed as a percent of the initial activity (i.e., activity in 100-mM SDS at room temperature, before heating at 90 °C). An exponential ($y = Ae^{-x/k}$) was fit to both data points and yielded time constants (k) of 94 ± 4 minutes for BLA-Ac(~17) ($R^2 = 0.969$) and 17 ± 1 minute for BLA-Ac(0) ($R^2 = 0.946$); BLA-Ac(~17) had a 5.5-fold greater half-life than BLA-Ac(0). Numerical integration yielded a total product ratio (BLA-Ac(~17): BLA-Ac (0)) of 5.2. Error bars in (A) represent the standard deviation of activity (%) from the mean ($N = 7$). Values <5% of the initial activity were ignored because of uncertain reliability. (B) A comparison of the protective effects of lysine acetylation against BLA inactivation in anionic (SDS), cationic (DTAB), and nonionic (Triton X-100) surfactants. Error bars represent the standard deviation of activity values from the mean ($N = 7$). Enzymatic activity of BLA was measured using the same protocol described for (A). Adapted from Shaw and Whitesides (Shaw et al., 2008).

XIII. Acetylation Increases the Half-Life of BLA in Hot Solutions of Sodium Dodecyl Sulfate

The major chemical ingredients of typical commercial or household detergent include the anionic surfactants sodium dodecyl sulfate (SDS; also known as sodium lauryl sulfate, SLS) and sodium lauryl ether sulfate (SLES); the neutral surfactant Triton X-100 (4-[1,1,3,3-tetramethylbutyl] phenyl-polyethylene glycol) is another common component. Cationic surfactants are not commonly used because they are toxic to humans and marine organisms. A typical formulation of liquid detergent will have a concentration of surfactant between 5% and 20% by weight (Saeki et al., 2007). Any enzyme that is used as an additive to detergent will therefore have to retain or regain activity after being stored in the high concentration of surfactant (Shaw et al., 2008). The requirements of the enzyme to be active are even greater in commercial applications where hot solutions of surfactant and enzyme (at 60–70 °C) are reused.

In order to determine if the acetylated BLA protein is protected from irreversible inactivation by surfactant and heat, more so than the unmodified protein, the enzymatic activity of BLA-Ac(0) and BLA-Ac(~17) was measured after exposure to hot solutions of SDS (100 mM, pH 8.5, 90 °C). Each protein was incubated under these conditions for up to 9 hours and the activity was measured *after* an aliquot was removed, cooled to room temperature, and then diluted (1:181 v/v) into the Ceralpha assay buffer. Neither BLA-Ac(~17) nor BLA-Ac(0) is enzymatically active under the harsh preassay conditions (100-mM SDS (3% SDS), 90 °C).

After a 1 hour preassay incubation at room temperature, both proteins were observed to have similar enzymatic activity (Fig. 7B). Upon preassay heating to 90 °C, however, the BLA-Ac(0) protein was inactivated more rapidly than BLA-Ac(~17); the activity of BLA-Ac(0) decreased by 60% after 20 minutes and by 95% after 60 minutes at 90 °C, 100 mM-SDS (Fig. 8A). In contrast, the BLA-Ac(~17) protein was more resistant to inactivation; only 30% of its activity was lost after 20 minutes at 90 °C, 100 mM-SDS, and only 60% of the activity was lost after 60 minutes.

The inactivation plots in Fig. 8A were fit with a single exponential function ($y = Ae^{-x/k}$) in order to calculate the differences in half-lives (k) of the two preparations. These calculations used only those data points with values that were >5% of the initial activity of each preparation because values that are <5% of the initial activity are difficult to measure.

The time constant, *k* (e.g., the half-life of the enzyme), was calculated to be 94 ± 4 minutes for BLA-Ac(~17) and 17.1 ± 1 minute for BLA-Ac(0); the acetylation of BLA, therefore, increased the half-life of the enzyme by a factor of 5.5. The inactivation plots for BLA-Ac(0) and BLA-Ac(~17) in Fig. 8A were also numerically integrated in order to quantify the degree to which acetylation of lysine-ε-NH_3^+ increases the chemical yield of the enzyme. The total product ratio (e.g., the ratio of integrals of the inactivation plot of BLA-Ac(~17) to BLA-Ac(0)) was calculated to be 5.2; the highly acetylated BLA protein therefore produced 5.2 times more product throughout the 9-hour experiment than BLA-Ac(0).

XIV. Why Does the Acetylation of Lysine Residues Increase the Half-Life of α-Amylase in Solutions of Surfactant?

The reasons why the acetylation of lysine residues in BLA protects the enzyme from inactivation after storage in hot surfactant solutions cannot be explained *entirely* by the lower rate of aggregation or precipitation of BLA-Ac(~17) compared to BLA-Ac(0). It is hypothesized that at least *two* different electrostatic effects are involved: a) the electrostatic repulsion between similarly charged BLA polypeptides (e.g., a decreased rate of aggregation of BLA) and b) electrostatic repulsion between the anionic surfactant and the negatively charged BLA polypeptide. Previous work from the Whitesides laboratory using protein charge ladders of bovine carbonic anhydrase II and capillary electrophoresis has shown that the acetylation of lysine residues can increase the kinetic stability of BCA II toward denaturation in SDS by a factor of 1,000. This stabilizing effect was attributed to an increase in electrostatic repulsion between the negatively charged surfactant and the more negatively charged BCA II protein that results from acetylation of lysine-ε-NH_3^+. Moreover, the stabilizing effect (kinetically) of acetylation against the unfolding of BCA II in SDS was observed even though acetylation also reduced the temperature of unfolding of BCA II (in solutions containing no surfactant) by approximately 20°C.

It is not surprising, therefore, that the increased resistance of BLA-Ac (~17) to undergo inactivation in hot surfactant depends, at least to some degree, upon the *charge* of the surfactant (Fig. 8). The more negatively charged BLA-Ac(~17) protein undergoes inactivation in surfactant more slowly than BLA-Ac(0)—even at temperatures approaching the boiling point of water—but only when the surfactant is neutral or negatively charged. Acetylation does *not* result in the protection to BLA from inactivation in the presence of a *cationic* surfactant (Fig. 8B). In the presence of the cationic surfactant DTAB (dodecyltrimethyl ammonium bromide), both BLA-Ac(~17) and BLA-Ac(0) were nearly completely inactivated after 60 minutes at 90 °C; both proteins retain <5% of their initial activity after 60 minutes. This result correlates with other studies that examined the binding of oppositely charged molecules and macromolecules with one another. Charged macromolecules and charged surfactants have been shown in numerous studies to bind more favorably—kinetically or thermodynamically—to proteins (and DNA) that have opposite charges (La Mesa, 2005; Liu and Guo, 2007). Cationic surfactant molecules might, therefore, bind to the highly negatively charged BLA-Ac(~17) protein (and to BLA-Ac(0)) more tightly than SDS and perhaps might remain bound after the 1:181 dilution in the Ceralpha assay medium. The binding of DTAB to unfolded BLA could inhibit the proper refolding of BLA, which would reduce the enzymatic activity during the assay.

The acetylation of lysine residues also protected BLA from irreversible inactivation in the presence of the neutral surfactant Triton X-100 at 90 °C. The acetylated BLA protein retained a greater amount of activity after an hour in Triton X-100 (1–100 mM) than in SDS (1–100 mM). It is reasonable to speculate that the BLA-Ac(~17) protein had a greater activity after 1 hour in Triton than in SDS or DTAB because the neutral surfactant binds less tightly to the protein than either the anionic or the cationic surfactant and does not, therefore, inhibit the proper refolding of the protein upon dilution into the assay solution.

XV. Identifying the Unmodified Lysine Residues in BLA-Ac(~17)

A mass spectrometric analysis of BLA-Ac(~17) suggested that not all lysine residues in BLA undergo acetylation to a similar extent and that a few lysine residues remain refractory to acetylation in the face of vast stoichiometric excesses of acetic anhydride. Solutions of BLA-Ac(~17) were proteolyzed with trypsin and also with elastase, and the resulting peptide mixture was subjected to a standard bottom-up protein identification protocol with LC/ESI-MS/MS. Elastase was used in addition to trypsin, the conventional proteolytic reagent used in bottom-up proteomics, because the acetylation of lysine can inhibit trypsin proteolysis. The results of this analysis are shown in Fig. 9; unmodified lysine residues are denoted with an asterisk; modified lysine residues are highlighted in magenta. Table I summarizes a comparison of the lysine residues that were found to be acetylated in BLA with some of the structural features of each lysine residue in BLA (i.e., burial from solvent, secondary structure, and ionic interactions according to X-ray crystallography).

This analysis revealed that 18 different lysine residues were acetylated in the BLA-Ac(~17) protein (Lys 70, 80, 88, 154, 176, 213, 234, 237, 251, 306, 315, 319, 370, 381, 383, 389, 392, and 436). Data regarding the acetylation of two lysine residues (Lys 254 and 276) were ambiguous. Five lysine residues were found to be void of modification (Lys 47, 76, 106, 180, and 344). Three lysine residues (Lys 23, 136, and 170) and the N-terminal alanine 1 were not contained in the sequence coverage and the acetylation of these residues could not be verified (~89% of the BLA sequence was identified with ESI-MS/MS).

What are the reasons for some lysine residues to be modified and some not modified? The burial of lysine residues must certainly be a factor; only one lysine residue is completely buried (Lys 47) and this lysine is one of the five that was not modified according to the analysis with MS/MS (Fig. 9 and Table I). Three lysine residues are also partially buried, and all three of these residues were observed to be unmodified (e.g., Lys 76, 106, and 344). Ionic interactions (i.e., salt-bridges) between lysine and nearby aspartate and glutamate residues do not appear, surprisingly, to prevent lysine residues from being modified. The X-ray crystal structure of BLA shows that 14 of the 28 lysine residues in BLA undergo ionic interactions with glutamate and/or aspartate residues; seven of these lysine residues were nevertheless found to be acetylated (Table I). There does appear, from the data shown in Table I, to be a correlation between the secondary structure of a lysine residue in BLA and its susceptibility to acetylation by acetic anhydride. For example, lysine residues in BLA were modified to a similar extent regardless of whether they were contained within α-helices or β-strands, but lysine residues that are contained within loops were more protected from acetylation: four of the five lysine residues that were not acetylated are contained in loops according to the X-ray crystal structure of BLA. It is also worth-mentioning that only one BLA lysine residue (Lys 234) is strictly

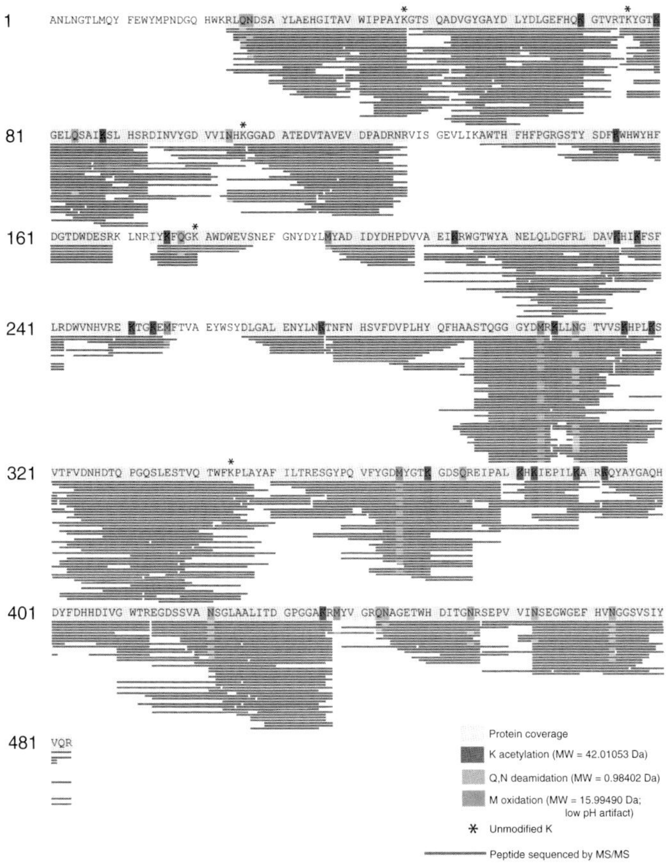

FIG. 9. Identification of acetylated lysine residues in a sample of "BLA-Ac(~17)" with a bottom-up proteomic approach (LC/ESI-MS/MS). Eighteen different lysine residues were found to have a 42-Da modification (e.g., Lys 70, 80, 88, 154, 176, 213, 234, 237, 251, 306, 315, 319, 370, 381, 383, 389, 392, and 436). Mass spectrometric data regarding the acetylation of two lysine residues (e.g., Lys 254 and Lys 276) were ambiguous. Five lysine residues were found to have no modifications (e.g., Lys 47, 76, 106, 180, and 344; denoted with an asterisk (*)). Three lysine residues (e.g., Lys 23, 136, and 170) and the N-terminal alanine 1 were not contained in the covered sequence and the acetylation of these residues can not be verified. BLA-Ac (~17) was proteolyzed with trypsin and elastase and the digest was analyzed with LC/ESI-MS/MS. Approximately 89% of the amino acid sequence of BLA was accounted for. Four methionine residues (e.g., Met 197, 256, 366, and 438) were oxidized (methionine oxidation is usually observed in bottom-up proteomic analyses). Adapted from Shaw and Whitesides (Shaw et al., 2008). (See color plate 11)

conserved across species, and this residue is crucial for BLA activity. Lysine 234 was, surprisingly, found to be modified without the loss of activity of the BLA protein. From the existing data it is not, however, possible to calculate the percentage of BLA that contained a modified Lys 234 in the BLA-Ac(~17) preparation. We cannot, therefore, ignore the possibility that only a small fraction of BLA molecules in the preparation contained an acetyl modification at Lys 234.

XVI. Conclusion and Outlook: *Charge Boosting* Drugs

Experimental evidence from the fields of genetics, protein biophysics, and protein engineering is accumulating in support of the *electrostatic hypothesis*—that the *net* electrostatic charge of a protein has a profound influence on the rate of aggregation and the toxicity of the protein. The net charge of a protein is an important physical property that must be considered when attempting to understand the rates of aggregation of proteins.

It is postulated that a pathogenic amino acid substitution that promotes the aggregation of its gene product, and triggers the pathogenesis of a protein "misfolding" disease, might operate—in some cases—by reducing the net charge of the pathogenic protein without necessarily altering the conformational stability, native structure, or the kinetics of folding of the protein. Does this electrostatic hypothesis provide any insight into new strategies for treating protein aggregation diseases? A few research groups have already shown that small molecules can be used to decrease the rate of aggregation of a pathogenic protein (Berg, 2003; Dumoulin et al., 2003; Arkin and Wells, 2004; Johnson et al., 2005; Wiseman et al., 2005; Johnson et al., 2008a, 2008b). Several small molecules have been found to bind to, and stabilize, globular proteins that cause amyloidoses (Johnson et al., 2005; Wiseman et al., 2005; Johnson et al., 2008a, 2008b) (e.g., transthyretin and lysozyme) and ALS (Ray and Lansbury, 2004; Ray et al., 2005) (e.g., SOD1), thus slowing the rate of aggregation. Many of these small molecules are electrically neutral and decrease the rate of aggregation of the protein by increasing the kinetic stability of the protein, that is, by decreasing the population of unfolded or partially unfolded states. The electrostatic hypothesis, however, suggests that the rate of aggregation of proteins could also be decreased (perhaps more easily) by the binding of small molecules that increase the net charge of the polypeptide. It seems reasonable to predict that a new class of drugs can be developed that operate as "charge boosters" and bind to target proteins and increase their net surface charge. Also, small molecule inhibitors of aggregation that might already be identified as a potential drug therapy might be improved by introducing an appropriately charged functional group to the molecule. One advantage of designing compounds that inhibit protein aggregation by increasing the net electrostatic charge of a protein, as opposed to increasing the stability of the native state, is that the former might be, practically speaking, much easier to do. The intermolecular electrostatic interactions that are involved in protein aggregation are *long-range* interactions (Warwicker, 1986; Daopin et al., 1991; Xu and Yeung, 1998; Caravella et al., 1999; Walser et al., 2001; del Alamo and Mateu, 2005; Gitlin et al., 2006a; Robertson et al., 2008) and the binding of a small, charged molecule to *any* site on the protein could increase the electrostatic repulsion between similarly charged proteins that are prone to aggregate with one another. Surprisingly, and in spite of the recognized role that net charge plays in protein aggregation, there have been no reported efforts yet to discover or design small molecules that inhibit protein aggregation by increasing the net charge of the target protein.

Acknowledgment

The authors would like to thank George M. Whitesides for his careful reading, and critique, of this chapter.

References

Abernethy, J. L., Steinman, H. M., Hill, R. L., (1974). Bovine erythrocyte superoxide dismutase. Subunit structure and sequence location of the intrasubunit disulfide bond. *J. Biol. Chem.* **249**, 7339–7347.

Antonyuk, S., Elam, J. S., Hough, M. A., Strange, R. W., Doucette, P. A., Rodriguez, J. A., et al., (2005). Structural consequences of the familial amyotrophic lateral sclerosis SOD1 mutant his46arg. *Protein Sci.* **14**, 1201–1213.

Arkin, M. R., Wells, J. A., (2004). Small-molecule inhibitors of protein-protein interactions: progressing towards the dream. *Nat. Rev. Drug Discov.* **3**, 301–317.

Assfalg, M., Banci, L., Bertini, I., Turano, P., Vasos, P. R., (2003). Superoxide dismutase folding/unfolding pathway: role of the metal ions in modulating structural and dynamical features. *J. Mol. Biol.* **330**, 145–158.

Baker, N. A., Sept, D., Joseph, S., Holst, M. J., McCammon, J. A., (2001). Electrostatics of nanosystems: application to microtubules and the ribosome. *Proc. Natl. Acad. Sci. U.S.A.* **98**, 10037–10041.

Banci, L., Bertini, I., Cramaro, F., Del Conte, R., Viezzoli, M. S., (2003). Solution structure of apo Cu, Zn superoxide dismutase: role of metal ions in protein folding. *Biochemistry* **42**, 9543–9553.

Banci, L., Bertini, I., D'Amelio, N., Gaggelli, E., Libralesso, E., Matecko, I., et al., (2005). Fully metallated S134N cu,zn-superoxide dismutase displays abnormal mobility and intermolecular contacts in solution. *J. Biol. Chem.* **280**, 35815–35821.

Basak, S. K., Ladisch, M. R., (1995). Correlation of electrophoretic mobilities of proteins and peptides with their physicochemical properties. *Anal. Biochem.* **226**, 51–58.

Berg, T., (2003). Modulation of protein-protein interactions with small organic molecules. *Angew. Chem. Int. Ed. Engl.* **42**, 2462–2481.

Caravella, J. A., Carbeck, J. D., Duffy, D. C., Whitesides, G. M., Tidor, B., (1999). Long-range electrostatic contributions to protein-ligand binding estimated using protein charge ladders, affinity capillary electrophoresis, and continuum electrostatic theory. *J. Am. Chem. Soc.* **121**, 4340–4347.

Cattaneo, E., Zuccato, C., Tartari, M., (2005). Normal huntingtin function: an alternative approach to Huntington's disease. *Nat. Rev. Neurosci.* **6**, 919–930.

Chiti, F., Calamai, M., Taddei, N., Stefani, M., Ramponi, G., Dobson, C. M., (2002). Studies of the aggregation of mutant proteins *in vitro* provide insights into the genetics of amyloid diseases. *Proc. Natl. Acad. Sci. U.S.A.* **99** (Suppl. 4), 16419–16426.

Chiti, F., Dobson, C. M., (2006). Protein misfolding, functional amyloid, and human disease. *Annu. Rev. Biochem.* **75**, 333–366.

Chiti, F., Stefani, M., Taddei, N., Ramponi, G., Dobson, C. M., (2003). Rationalization of the effects of mutations on peptide and protein aggregation rates. *Nature* **424**, 805–808.

Colton, I. J., Anderson, J. R., Gao, J. M., Chapman, R. G., Isaacs, L., Whitesides, G. M., (1997). Formation of protein charge ladders by acylation of amino groups on proteins. *J. Am. Chem. Soc.* **119**, 12701–12709.

Daopin, S., Soderlind, E., Baase, W. A., Wozniak, J. A., Sauer, U., Matthews, B. W., (1991). Cumulative site-directed charge-change replacements in bacteriophage-T4 lysozyme suggest that long-range electrostatic interactions contribute little to protein stability. *J. Mol. Biol.* **221**, 873–887.

Declerck, N., Machius, M., Joyet, P., Wiegand, G., Huber, R., Gaillardin, C., (2003). Hyperthermostabilization of *Bacillus licheniformis* alpha-amylase and modulation of its stability over a 50 degrees C temperature range. *Protein Eng.* **16**, 287–293.

del Alamo, M., Mateu, M. G., (2005). Electrostatic repulsion, compensatory mutations, and long-range non-additive effects at the dimerization interface of the HIV capsid protein. *J. Mol. Biol.* **345**, 893–906.

DiDonato, M., Craig, L., Huff, M. E., Thayer, M. M., Cardoso, R. M., Kassmann, C.J., et al., (2003). ALS mutants of human superoxide dismutase form fibrous aggregates via framework destabilization. *J. Mol. Biol.* **332**, 601–615.

Dumoulin, M., Last, A. M., Desmyter, A., Decanniere, K., Canet, D., Larsson, G., et al., (2003). A camelid antibody fragment inhibits the formation of amyloid fibrils by human lysozyme. *Nature* **424**, 783–788.

Dyson, H. J., Wright, P. E., (2005). Intrinsically unstructured proteins and their functions. *Nat. Rev. Mol. Cell Biol.* **6**, 197–208.

Edgcomb, S. P., Murphy, K. P., (2002). Variability in the pKa of histidine side-chains correlates with burial within proteins. *Proteins* **49**, 1–6.

Elam, J. S., Taylor, A. B., Strange, R., Antonyuk, S., Doucette, P. A., Rodriguez, J. A., et al., (2003). Amyloid-like filaments and water-filled nanotubes formed by SOD1 mutant proteins linked to familial ALS. *Nat. Struct. Biol.* **10**, 461–467.

Forman, H. J., Fridovich, I., (1973). On the stability of bovine superoxide dismutase. The effects of metals. *J. Biol. Chem.* **248**, 2645–2649.

Gao, J. M., Whitesides, G. M., (1997). Using protein charge ladders to estimate the effective charges and molecular weights of proteins in solution. *Anal. Chem.* **69**, 575–580.

Garcia-Moreno, B., Dwyer, J. J., Gittis, A. G., Lattman, E. E., Spencer, D. S., Stites, W. E., (1997). Experimental measurement of the effective dielectric in the hydrophobic core of a protein. *Biophys. Chem.* **64**, 211–224.

Gitlin, I., Carbeck, J. D., Whitesides, G. M., (2006a). Why are proteins charged? Networks of charge-charge interactions in proteins measured by charge ladders and capillary electrophoresis. *Angew. Chem. Int. Ed. Engl.* **45**, 3022–3060.

Gitlin, I., Gudiksen, K. L., Whitesides, G. M., (2006b). Effects of surface charge on denaturation of bovine carbonic anhydrase. *Chembiochem* **7**, 1241–1250.

Gitlin, I., Gudiksen, K. L., Whitesides, G. M., (2006c). Peracetylated bovine carbonic anhydrase (BCA-ac18) is kinetically more stable than native BCA to sodium dodecyl sulfate. *J. Phys. Chem. B* **110**, 2372–2377.

Grossman, P. D., (1992). Capillary electrophoresis: theory and practice. Academic Press, San Diego, CA, pp. 111–132.

Gudiksen, K. L., Gitlin, I., Moustakas, D. T., Whitesides, G. M., (2006). Increasing the net charge and decreasing the hydrophobicity of bovine carbonic anhydrase decreases the rate of denaturation with sodium dodecyl sulfate. *Biophys. J.* **91**, 298–310.

Gudiksen, K. L., Gitlin, I., Yang, J., Urbach, A. R., Moustakas, D. T., Whitesides, G. M., (2005). Eliminating positively charged lysine epsilon-NH3+ groups on the surface of carbonic anhydrase has no significant influence on its folding from sodium dodecyl sulfate. *J. Am. Chem. Soc.* **127**, 4707–4714.

Gurney, M. E., Pu, H., Chiu, A. Y., Dal Canto, M. C., Polchow, C. Y., Alexander, D. D., et al., (1994). Motor neuron degeneration in mice that express a human Cu, Zn superoxide dismutase mutation. *Science* **264**, 1772–1775.

Hansch, C., Coats, E., (1970). Alpha-chymotrypsin: a case study of substituent constants and regression analysis in enzymic structure–activity relationships. *J. Pharm. Sci.* **59**, 731–743.

Hansch, C., Steward, A. R., (1964). The use of substituent constants in the analysis of the structure–activity relationship in penicillin derivatives. *J. Med. Chem.* **7**, 691–694.

Harjes, P., Wanker, E. E., (2003). The hunt for huntingtin function: interaction partners tell many different stories. *Trends Biochem. Sci.* **28**, 425–433.

Hart, P. J., Liu, H., Pellegrini, M., Nersissian, A. M., Gralla, E. B., Valentine, J. S., et al., (1998). Subunit asymmetry in the three-dimensional structure of a human CuZnSOD mutant found in familial amyotrophic lateral sclerosis. *Protein Sci.* **7**, 545–555.

Hayward, L. J., Rodriguez, J. A., Kim, J. W., Tiwari, A., Goto, J. J., Cabelli, D. E., et al., (2002). Decreased metallation and activity in subsets of mutant superoxide dismutases associated with familial amyotrophic lateral sclerosis. *J. Biol. Chem.* **277**, 15923–15931.

Hough, M. A., Grossmann, J. G., Antonyuk, S. V., Strange, R. W., Doucette, P. A., Rodriguez, J. A., et al., (2004). Dimer destabilization in superoxide dismutase may result in disease-causing properties: structures of motor neuron disease mutants. *Proc. Natl. Acad. Sci. U.S.A.* **101**, 5976–5981.

Hwang, K. Y., Song, H. K., Chang, C., Lee, J., Lee, S. Y., Kim, K. K., et al., (1997). Crystal structure of thermostable alpha-amylase from *Bacillus licheniformis* refined at 1.7 A resolution. *Mol. Cells* **7**, 251–258.

Jain, R., Hamilton, A. D., (2002). Designing protein denaturants: synthetic agents induce cytochrome c unfolding at low concentrations and stoichiometries. *Angew. Chem. Int. Ed. Engl.* **41**, 641–643.

Johnson, S. M., Connelly, S., Wilson, I. A., Kelly, J. W., (2008a). Biochemical and structural evaluation of highly selective 2-arylbenzoxazole-based transthyretin amyloidogenesis inhibitors. *J. Med. Chem.* **51**, 260–270.

Johnson, S. M., Connelly, S., Wilson, I. A., Kelly, J.W., (2008b). Toward optimization of the linker substructure common to transthyretin amyloidogenesis inhibitors using biochemical and structural studies. *J. Med. Chem.* **51**, 6348–6358.

Johnson, S. M., Wiseman, R. L., Sekijima, Y., Green, N. S., Adamski-Werner, S. L., Kelly, J. W., (2005). Native state kinetic stabilization as a strategy to ameliorate protein misfolding diseases: a focus on the transthyretin amyloidoses. *Acc. Chem. Res.* **38**, 911–921.

Keele, B. B. J., McCord, J. M., Fridovich, I., (1971). Further characterization of bovine superoxide dismutase and its isolation from bovine heart. *J. Biol. Chem.* **246**, 2875–2880.

Kirk, O., Borchert, T. V., Fuglsang, C. C., (2002). Industrial enzyme applications. *Curr. Opin. Biotechnol.* **13**, 345–351.

Koo, E., (2006). Amyloid beta-protein and amyloid precursor protein: how "seminal" are they in Alzheimer's disease? *J. Neurochem.* **98**, 12–12.

Krishnamurthy, V. M., Kaufman, G. K., Urbach, A. R., Gitlin, I., Gudiksen, K. L., Weibel, D.B., et al., (2008). Carbonic anhydrase as a model for biophysical and physical-organic studies of proteins and protein-ligand binding. *Chem. Rev.* **108**, 946–1051.

La Mesa, C., (2005). Polymer-surfactant and protein-surfactant interactions. *J. Colloid Interface Sci.* **286**, 148–157.

Lawrence, M. S., Phillips, K. J., Liu, D. R., (2007). Supercharging proteins can impart unusual resilience. *J. Am. Chem. Soc.* **129**, 10110–10112.

Lee, B., Vasmatzis, G., (1997). Stabilization of protein structures. *Curr. Opin. Biotechnol.* **8**, 423–428.

Lindberg, M. J., Bystrom, R., Boknas, N., Andersen, P. M., Oliveberg, M., (2005). Systematically perturbed folding patterns of amyotrophic lateral sclerosis (ALS)-associated SOD1 mutants. *Proc. Natl. Acad. Sci. U.S.A.* **102**, 9754–9759.

Lindberg, M. J., Normark, J., Holmgren, A., Oliveberg, M., (2004). Folding of human superoxide dismutase: disulfide reduction prevents dimerization and produces marginally stable monomers. *Proc. Natl. Acad. Sci. U.S.A.* **101**, 15893–15898.

Liu, Y., Guo, R. ((2007)). Interaction between casein and the oppositely charged surfactant. *Biomacromolecules.* **8**, 2902–2908

Lynd, L. R., Laser, M. S., Bransby, D., Dale, B. E., Davison, B., Hamilton, R., et al., (2008). How biotech can transform biofuels. *Nat. Biotechnol.* **26**, 169–172.

Machius, M., Declerck, N., Huber, R., Wiegand, G., (1998). Activation of *Bacillus licheniformis* alpha-amylase through a disorder- ->order transition of the substrate-binding site mediated by a calcium-sodium-calcium metal triad. *Structure* **6**, 281–292.

Machius, M., Declerck, N., Huber, R., Wiegand, G., (2003). Kinetic stabilization of *Bacillus licheniformis* alpha-amylase through introduction of hydrophobic residues at the surface. *J. Biol. Chem.* **278**, 11546–11553.

Machius, M., Wiegand, G., Huber, R., (1995). Crystal structure of calcium-depleted *Bacillus licheniformis* alpha-amylase at 2.2 A resolution. *J. Mol. Biol.* **246**, 545–559.

Matthews, B. W., (1993). Structural and genetic analysis of protein stability. *Annu. Rev. Biochem.* **62**, 139–160.

McCleary, B. V., McNally, M., Monaghan, D., Mugford, D. C., (2002). Measurement of alpha-amylase activity in white wheat flour, milled malt, and microbial enzyme preparations, using the Ceralpha assay: collaborative study. *J. AOAC Int.* **85**, 1096–1102.

McCord, J. M., Fridovich, I., (1969). Superoxide dismutase. An enzymic function for erythrocuprein (hemocuprein). *J. Biol. Chem.* **244**, 6049–6055.

Menon, M. K., Zydney, A. L., (2000). Determination of effective protein charge by capillary electrophoresis: effects of charge regulation in the analysis of charge ladders. *Anal. Chem.* **72**, 5714–5717.

Miller, S., Janin, J., Lesk, A. M., Chothia, C., (1987a). Interior and surface of monomeric proteins. *J. Mol. Biol.* **196**, 641–656.

Miller, S., Lesk, A. M., Janin, J., Chothia, C., (1987b). The accessible surface area and stability of oligomeric proteins. *Nature* **328**, 834–836.

Millhauser, G. L., (2007). Copper and the prion protein: methods, structures, function, and disease. *Annu. Rev. Phys. Chem.* **58**, 299–320.

Pauling, L., (1993). How my interest in proteins developed. *Protein Sci.* **2**, 1060–1063.

Potter, S. Z., Zhu, H., Shaw, B. F., Rodriguez, J. A., Doucette, P. A., Sohn, S. H., et al., (2007). Binding of a single zinc ion to one subunit of copper-zinc superoxide dismutase apoprotein substantially influences the structure and stability of the entire homodimeric protein. *J. Am. Chem. Soc.* **129**, 4575–4583.

Ray, S. S., Lansbury, P. T. Jr., (2004). A possible therapeutic target for Lou Gehrig's disease. *Proc. Natl. Acad. Sci. U.S.A.* **101**, 5701–5702.

Ray, S. S., Nowak, R. J., Brown, R. H. Jr., Lansbury, P. T. Jr., (2005). Small-molecule-mediated stabilization of familial amyotrophic lateral sclerosis-linked superoxide dismutase mutants against unfolding and aggregation. *Proc. Natl. Acad. Sci. U.S.A.* **102**, 3639–3644.

Richardson, J., Thomas, K. A., Rubin, B. H., Richardson, D. C., (1975). Crystal structure of bovine Cu, Zn superoxide dismutase at 3 A resolution: chain tracing and metal ligands. *Proc. Natl. Acad. Sci. U.S.A.* **72**, 1349–1353.

Rickard, E. C., Strohl, M. M., Nielsen, R. G., (1991). Correlation of electrophoretic mobilities from capillary electrophoresis with physicochemical properties of proteins and peptides. *Anal. Biochem.* **197**, 197–207.

Roberts, B. R., Tainer, J. A., Getzoff, E. D., Malencik, D. A., Anderson, S. R., Bomben, V. C., et al., (2007). Structural characterization of zinc-deficient human superoxide dismutase and implications for ALS. *J. Mol. Biol.* **373**, 877–890.

Robertson, A., Luttmann, E., Pande, V. S., (2008). Effects of long-range electrostatic forces on simulated protein folding kinetics. *J. Comput. Chem.* **29**, 694–700.

Rodriguez, J. A., Shaw, B. F., Durazo, A., Sohn, S. H., Doucette, P. A., Nersissian, A. M., et al., (2005). Destabilization of apoprotein is insufficient to explain Cu,Zn-superoxide dismutase-linked ALS pathogenesis. *Proc. Natl. Acad. Sci. U.S.A.* **102**, 10516–10521.

Rodriguez, J. A., Valentine, J. S., Eggers, D. K., Roe, J. A., Tiwari, A., Brown, R. H. Jr., et al. (2002). Familial amyotrophic lateral sclerosis-associated mutations decrease the thermal stability of distinctly metallated species of human copper/zinc superoxide dismutase. *J. Biol. Chem.* **277**, 15932–15937.

Rosen, D. R., Siddique, T., Patterson, D., Figlewicz, D. A., Sapp, P., Hentati, A., et al., (1993). Mutations in Cu/Zn superoxide dismutase gene are associated with familial amyotrophic lateral sclerosis. *Nature* **362**, 59–62.

Rumfeldt, J. A., Lepock, J. R., Meiering, E. M., (2009). Unfolding and folding kinetics of amyotrophic lateral sclerosis-associated mutant Cu,Zn superoxide dismutases. *J. Mol. Biol.* **385**, 278–298.

Saeki, K., Ozaki, K., Kobayashi, T., Ito, S., (2007). Detergent alkaline proteases: enzymatic properties, genes, and crystal structures. *J. Biosci. Bioeng.* **103**, 501–508.

Sandelin, E., Nordlund, A., Andersen, P. M., Marklund, S. S., Oliveberg, M., (2007). Amyotrophic lateral sclerosis-associated copper/zinc superoxide dismutase mutations preferentially reduce the repulsive charge of the proteins. *J. Biol. Chem.* **282**, 21230–21236.

Shaw, B. F., Durazo, A., Nersissian, A. M., Whitelegge, J. P., Faull, K. F., Valentine, J. S., (2006). Local unfolding in a destabilized, pathogenic variant of superoxide dismutase 1 observed with H/D exchange and mass spectrometry. *J. Biol. Chem.* **281**, 18167–18176.

Shaw, B. F., Schneider, G.F., Bilgicer, B., Kaufman, G. K., Neveu, J. M., Lane, W. S., et al., (2008). Lysine acetylation can generate highly charged enzymes with increased resistance toward irreversible inactivation. *Protein Sci.* **17**, 1446–1455.

Shaw, B. F., Valentine, J. S., (2007). How do ALS-associated mutations in superoxide dismutase 1 promote aggregation of the protein? *Trends Biochem. Sci.* **32**, 78–85.

Shipp, E. L., Cantini, F., Bertini, I., Valentine, J. S., Banci, L., (2003). Dynamic properties of the G93A mutant of copper-zinc superoxide dismutase as detected by NMR spectroscopy: implications for the pathology of familial amyotrophic lateral sclerosis. *Biochemistry* **42**, 1890–1899.

Sticklen, M., (2006). Plant genetic engineering to improve biomass characteristics for biofuels. *Curr. Opin. Biotechnol.* **17**, 315–319.

Strange, R. W., Antonyuk, S., Hough, M. A., Doucette, P. A., Rodriguez, J. A., Hart, P. J., et al., (2003). The structure of holo and metal-deficient wild-type human Cu, Zn superoxide dismutase and its relevance to familial amyotrophic lateral sclerosis. *J. Mol. Biol.* **328**, 877–891.

Sugase, K., Dyson, H. J., Wright, P. E., (2007). Mechanism of coupled folding and binding of an intrinsically disordered protein. *Nature* **447**, 1021–1025.

Thurlkill, R. L., Grimsley, G. R., Scholtz, J. M., Pace, C. N., (2006). PK values of the ionizable groups of proteins. *Protein Sci.* **15**, 1214–1218.

Torigoe, K., Hasegawa, S., Numata, O., Yazaki, S., Matsunaga, M., Boku, N., et al., (2000). Influence of emission from rice straw burning on bronchial asthma in children. *Pediatr. Int.* **42**, 143–150.

Valentine, J. S., Doucette, P. A., Zittin Potter, S., (2005). Copper-zinc superoxide dismutase and amyotrophic lateral sclerosis. *Annu. Rev. Biochem.* **74**, 563–593.

Vekrellis, K., Rideout, H. J., Stefanis, L., (2004). Neurobiology of alpha-synulein. *Mol. Neurobiol.* **30**, 1–21.

Vendruscolo, M., Dobson, C. M., (2007). Chemical biology: more charges against aggregation. *Nature* **449**, 555.

Walser, R., Hunenberger, P. H., van Gunsteren, W. F., (2001). Comparison of different schemes to treat long-range electrostatic interactions in molecular dynamics simulations of a protein crystal. *Proteins: Struct. Funct. Bioinf.* **43**, 509–519.

Wang, Y. C., Ha, Y., (2004). The X-ray structure of an antiparallel dimer of the human amyloid precursor protein E2 domain. *Mol. Cell* **15**, 343–353.

Wang, J., Xu, G., Borchelt, D. R., (2002). High molecular weight complexes of mutant superoxide dismutase 1: age-dependent and tissue-specific accumulation. *Neurobiol. Dis.* **9**, 139–148.

Warwicker, J., (1986). Continuum dielectric modeling of the protein solvent system, and calculation of the long-range electrostatic-field of the enzyme phosphoglycerate mutase. *J. Theor. Biol.* **121**, 199–210.

Wiseman, R. L., Johnson, S. M., Kelker, M. S., Foss, T., Wilson, I. A., Kelly, J. W., (2005). Kinetic stabilization of an oligomeric protein by a single ligand binding event. *J. Am. Chem. Soc.* **127**, 5540–5551.

Xu, X. H. N., Yeung, E. S., (1998). Long-range electrostatic trapping of single-protein molecules at a liquid-solid interface. *Science* **281**, 1650–1653.

Zuccato, C., Conti, L., Reitano, E., Tartari, M., Cattaneo, E., (2005). The function of the neuronal proteins shc and huntingtin in stem cells and neurons—pharmacologic exploitation for human brain diseases. *Stem Cell Biol. Dev. Plast.* **1049**, 39–50.

HISTONE ACETYLATION, ACETYLTRANSFERASES, AND ATAXIA—ALTERATION OF HISTONE ACETYLATION AND CHROMATIN DYNAMICS IS IMPLICATED IN THE PATHOGENESIS OF POLYGLUTAMINE-EXPANSION DISORDERS

By SHAUN D. McCULLOUGH AND PATRICK A. GRANT

Department of Biochemistry and Molecular Genetics, University of Virginia School of Medicine, Charlottesville, VA, USA

Abstract

Eukaryotic chromosomal DNA is packaged into nucleosomes to form a dynamic structure known as chromatin. The compaction of DNA within chromatin poses a unique hindrance with regards to the accessibility of the DNA to enzymes involved in replication, transcriptional regulation, and repair. The physical structure and physiological activity of chromatin are regulated through a diverse set of posttranslational modifications, histone exchange, and structural remodeling. Of the covalent chromatin modifications, the acetylation of lysine residues within histone proteins by

ADVANCES IN PROTEIN CHEMISTRY AND STRUCTURAL BIOLOGY, Vol. 79
DOI: 10.1016/S1876-1623(10)79005-2

acetyltransferase enzymes, such as GCN5, is one of the most prevalent and important steps in the regulation of chromatin function. Alteration of histone acetyltransferase activity can easily result in the dysregulation of gene transcription and ultimately the onset of a disease state. Many transcription factors contain polyglutamine regions within their primary sequence. Mutations resulting in the elongation of these polyglutamine tracts are associated with a disease family known as the polyglutamine expansion disorders. Spinocerebellar ataxia type 7 (SCA7) is one of the nine diseases that are grouped in this family and is caused by polyglutamine expansion of the ataxin-7 protein, which is a component of the GCN5-containing human SAGA histone acetyltransferase complex. Mutation of ataxin-7 in this manner has been shown to disrupt the structural integrity of the SAGA complex and result in aberrant chromatin acetylation patterns at the promoters of genes involved in the normal function of tissues that are affected by the disease. The specific aspects of molecular pathology are not currently understood; however, studies carried out in laboratory systems ranging from the budding yeast Saccharomyces cerevisiae to transgenic mouse models and cultured human cells are poised to allow for the elucidation of disease mechanisms and subsequent therapeutic approaches.

I. Packaging and Accessibility of DNA in the Context of Chromatin

Despite its molecular simplicity, DNA encodes information that serves as the blueprint for the structure and function of every living organism. The relative abundance of DNA precludes the possibility of a simple mechanism for its packaging and accessibility within the eukaryotic nucleus. Within the nucleus, DNA is wrapped around nucleosomes, which allow for its compaction into the structure called chromatin. The nucleosome core particle binds 146bp of DNA and is composed of two histone H2A/H2B dimers that interact with a heterotetramer containing two copies each of histones H3 and H4 (Luger et al., 1997). Each of the histone proteins within the nucleosome contains alpha helical histone fold domains, which facilitate their interaction with adjacent histones within individual nucleosomes and DNA. Histone proteins also possess unstructured amino-(N-) and carboxy-terminal (C-terminal) tails that extend outside the nucleosome particle and serve as a substrate for the enzymatic activities of histone-modifying enzymes. Histones are extensively modified by a myriad of reactions; examples include acetylation, methylation, ubiquitination, SUMOylation, phosphorylation, and ADP-ribosylation. These modifications alter chromatin dynamics and influence histone–DNA interactions, as well as the recruitment and binding of protein factors to chromatin (reviewed in Taverna et al., 2007). The tight correlation between histone modification patterns and their subsequent regulation of chromatin activity led to the hypothesis by Strahl and Allis (2000) that posttranslational modification of histones encodes a language that is read and followed by chromatin-interacting proteins. These so-called chromatin-binding proteins interact with chromatin through several distinct domains, each of which has the ability to recognize distinct histone modifications and can be responsible for the targeting of multisubunit protein complexes to specific chromatin regions.

Histone acetylation is a posttranslational modification that performs two known functions on chromatin. First, acetylation neutralizes the positive charge of lysine residues within the histone tails thereby reducing their ability to form charge-based interactions with the negatively charged phosphodiester backbone of associated DNA. The acetylation of histones in general is associated with chromatin decondensation, although it is not necessarily causative. One exception is the acetylation of histone H4 on lysine 16, which has been shown to directly promote a decondensed chromatin state (Shogren-Knaak et al., 2006). The decondensation of chromatin is associated with the formation of a euchromatin state, which increases the accessibility of chromatin to transcription factors. In contrast, chromatin condensation leads to the formation of heterochromatin, which is transcriptionally silent. The second known function of histone acetylation is to provide a platform for the recruitment of transcription factors through acetyl–lysine interacting motifs, known as bromodomains (Dhalluin et al., 1999; Jacobson et al., 2000).

Histone acetylation is accomplished by a family of enzymes known as histone acetyltransferases (HATs), which were originally divided into two groups based on their subcellular localization, the nuclear HATs (Type A) and cytoplasmic HATs (Type B). The activity and substrate recognition of many HATs is typically regulated by their association with other factors in the context of multiprotein complexes (reviewed in Grant and Berger, 1999; Roth et al., 2001; Torok and Grant, 2004). These protein complexes often contain smaller subcomplexes, termed "modules," each of which contribute a discrete activity to the greater complex. The modular nature of these multiprotein complexes allows them to couple multiple histone modification activities to effectively regulate chromatin accessibility and function. While not always required, the substrate recognition and activity level of subcomplex modules can be dependent on their association with the parent complex. Overall, histone acetylation serves as a significant regulatory mechanism governing chromatin accessibility and providing a substrate for the association of factors with DNA in chromatin.

II. Gcn5 is a Histone Acetyltransferase that Functions in the Context of SAGA and its Related Complexes in Eukaryotes

The general control of amino acid biosynthesis 5 (Gcn5) protein in budding yeast was originally identified as a transcription factor involved in the cellular response to amino acid starvation and as a factor required for the optimal activity of other transcriptional activators (Georgakopoulos and Thireos, 1992). Sequence comparison of the p55 protein of *Tetrahymena*, the first confirmed Type-A HAT protein (Brownell and Allis, 1995), indicated significant similarity to the yeast Gcn5 protein and further analysis demonstrated that purified, recombinant yeast Gcn5 was also able to acetylate free core histones *in vitro* (Brownell et al., 1996). Subsequent characterization of recombinant Gcn5 demonstrated its ability to catalyze the acetylation of both histone H3 at lysine 9 and 14 and histone H4 at lysine 8 and 16. Despite its ability to acetylate lysine residues on either histone H3 or H4, Gcn5 exhibits a clear preference for histone H3 as an acetylation substrate. Within histone H3, lysine 14 is preferentially acetylated *in vitro* while lysine 9 appears to be the favored target *in vivo* (Kuo et al., 1996). Despite its documented ability to acetylate free core histones, recombinant Gcn5 did not appear to be capable of acetylating histones within the nucleosome core particle, the context in which histones exist within chromatin. The lack of nucleosomal acetyltransferase activity suggested that other factors were required to regulate either the recognition of histone substrates within the nucleosome or HAT activity of the enzyme in this altered context.

Early studies involving the yeast Gcn5 protein demonstrated its ability to functionally interact both *in vitro* and *in vivo* with two adaptor proteins, Ada2 and Ada3, to form a trimeric complex with the ability to promote transcriptional activation (Marcus et al., 1994, Horiuchi et al., 1995; Candau et al., 1996). While it was subsequently shown by Balasubramanian et al. (2002) that the heterotrimeric complex formed by Gcn5, Ada2, and Ada3 had the ability to acetylate nucleosomal histones, its native association with other polypeptides to form larger chromatin-modifying complexes had not yet been explored. The search for these complexes led to an unbiased interrogation of multiprotein complexes possessing HAT activity through a multiple step, native chromatographic purification from yeast extracts. This exploration ultimately resulted in the identification of four distinct complexes with the ability to acetylate nucleosomal histones (Grant et al., 1997). Two of these four complexes were shown to contain Gcn5 and the adaptor protein Ada2, suggesting that they possessed the necessary catalytic activity and targeting ability to function as multisubunit acetyltransferase complexes. Subsequent characterization of the highest molecular weight HAT complex, estimated at 1.8 megadaltons, identified several additional subunits, including several members of the suppressor of Ty (Spt) protein family, Spt3, Spt7, and Spt20. The discovery of these subunits in association with the high molecular weight HAT complex led its naming as the Spt-Ada-Gcn5-acetyltransferase (SAGA) complex. During initial characterization of the yeast SAGA complex it was shown that deletion of Gcn5, Ada2, or Ada3 resulted in loss of HAT activity, indicating that Gcn5 was the sole acetyltransferase responsible for SAGA-mediated histone acetylation. Additionally, these experiments demonstrated that the adaptor proteins Ada2 and Ada3 were required for mediating the activity of Gcn5 in a physiologically relevant context. While Gcn5, Ada2, and Ada3 were not required for the overall integrity of the complex, they were shown to be critical for its ability to recognize and acetylate nucleosomal substrates and defined the Gcn5-Ada2-Ada3 heterotrimer as the catalytic core of the SAGA complex (Grant et al., 1997). It was later determined that there was a second Gcn5-containing protein complex in yeast that functioned in stress-response pathways and shared a similar subunit composition and histone substrate specificity with SAGA (Pray-Grant et al., 2002; Sterner et al., 2002). Due to its similarity to SAGA, this second Gcn5-containing complex was named the SAGA-like (SLIK) complex. Both the SAGA and SLIK complexes serve as examples of how histone acetylation by Gcn5 is controlled through its interactions with accessory factors in the context of multiprotein chromatin-modifying complexes. The composition and functions of these yeast acetyltransferase complexes has served as the basis for the identification and characterization of similar complexes in higher eukaryotes.

Unlike yeast, metazoans possess two GCN5 isoforms that are believed to arise from the alternative splicing of a single gene product (Smith et al., 1998). The lower molecular weight isoform (GCN5-S) contains both a HAT domain and a single bromodomain and is similar in size and function to the single form of Gcn5 found in yeast. The second, higher molecular weight isoform (GCN5-L) has been shown to contain an additional N-terminal domain that shows high similarity to the N-terminal domain of the GCN5 homolog p300/CREB-binding protein-associated factor, known as PCAF (Xu et al., 1998). Additionally, the vertebrate-specific histone acetyltransferase PCAF, identified as a paralog of human GCN5 (Yang et al., 1996), is also a constituent of a multisubunit HAT complex.

GCN5 isoforms serve as the catalytic acetyltransferases in several multiprotein complexes in human cells, including the TATA-binding protein-free TAF complex (TFTC), ADA-two-a-containing complex (ATAC), and the SPT3-TAF_{II}31-GCN5-L acetylase (STAGA) complex (reviewed in Nagy and Tora, 2007). The human STAGA complex was purified from cultured cells shortly after the identification and characterization of the homologous yeast SAGA complex (Martinez et al., 1998). The STAGA complex preferentially incorporates GCN5-L over both its homolog PCAF and the smaller GCN5 isoform and favors histone H3 as a substrate for acetylation compared to other histone proteins (Ogryzko et al., 1998; Martinez et al., 2001). Similar to yeast SAGA, the human STAGA complex contains human homologs of Ada2 and Ada3, the components of the GCN5 catalytic trimer, and has the ability to activate transcription from *in vitro*-assembled chromatin templates (Martinez et al., 2001). The subunit compositions of the yeast SAGA and human STAGA complexes are well conserved and demonstrate the ability to regulate gene expression through several different functions, which will be described below.

III. Substrate Targeting of Gcn5 Histone Acetyltransferase Activity is Regulated by Interactions with Adaptor Proteins

The control of substrate specificity and catalytic function by association with accessory proteins is a recurring theme in enzyme regulation throughout the cell. Proteins that exhibit HAT activity in the absence of accessory proteins often demonstrate the ability to target multiple lysine residues within individual, or multiple, histones. The degree of promiscuity demonstrated by acetyltransferases, in the absence of regulatory proteins, raises the question of how their specificities are controlled to ensure the appropriate spatiotemporal acetylation of histone targets to preserve the fidelity of transcriptional regulation. The exact mechanism of histone substrate targeting by these acetyltransferases is the subject of rigorous study as the answers to these and other related questions may facilitate the understanding of how gene expression in both normal and disease states is controlled by histone acetylation.

The monomeric Gcn5 protein in yeast acetylates both histone H3 and H4 when they are free in solution, but not when they are in the context of the nucleosome core particle *in vitro* (Xu et al., 1998). Microsequencing of histones acetylated by Gcn5 *in vitro* indicated the preferential acetylation of histone H3 at lysine 14 and histone H4 at lysine 8 and 16 (Kuo et al., 1996). These data provided the foundation for the characterization of HAT activity exerted by Gcn5; however, it was determined later that the substrate specificity of Gcn5 in the context of both the Gcn5-Ada2-Ada3 catalytic core and the greater SAGA complex varied from that observed for monomeric Gcn5 (Grant et al., 1999). The preferential targeting of lysine residues within the nucleosome core particle by Gcn5 was further studied and was shown to be dependent on its association with the adaptor proteins Ada2 and Ada3. In addition to the lysine specificity exhibited by monomeric Gcn5, the Gcn5-Ada2 heterodimer can acetylate lysine 18, although with a lower efficiency compared to lysine 14. Subsequent addition of Ada3, to complete the Gcn5-Ada2-Ada3 catalytic core, further alters target lysine specificity leading to acetylation of histone H3 on lysine 9, 14, 18, and 23 with variable efficiency. In the context of the entire SAGA complex, overall acetylation of lysine residues on histone H3 is decreased compared to the catalytic core; however, the substrate specificity and relative preference for specific lysine residues are similar between the two conditions (Balasubramanian et al., 2002). Similar to the SAGA complex, the lower molecular weight ADA complex efficiently acetylates histone H3; however, the ADA complex appears to demonstrate a more robust ability to acetylate histones H2B and H4 when compared to SAGA. The difference in substrate targeting of ADA coincides with the incorporation of a different complement of proteins into this functionally distinct, Gcn5-containing complex (Grant et al.,

1997, 1999; Sterner et al., 1999). These observations suggest the potential for Gcn5 to regulate transcription, or other chromatin-related activities, through multiple mechanisms based on protein–protein interactions within multisubunit complexes.

In contrast to yeast Gcn5, the larger GCN5 isoform found in higher eukaryotes can acetylate histones H3 and H4 in both the free and nucleosome-incorporated states *in vitro* without the presence of accessory proteins (Xu et al., 1998). The ability of free GCN5-L protein to acetylate nucleosomal histones in more complex organisms is presumably dependent on the presence of the additional N-terminal domain, which is absent from yeast Gcn5. As mentioned above, Ada2 is involved in facilitating the recognition of nucleosomal histone substrates by Gcn5 in yeast. During the course of evolution, the single Ada2 gene in yeast developed into two functionally distinct ADA2 paralogs in metazoans, ADA2-alpha (ADA2a) and ADA2-beta (ADA2b). Interestingly, GCN5 functionally interacts with both of these ADA2 isoforms in *Drosophila* (Muratoglu et al., 2003), mice (Candau et al., 1996), and humans (Barlev et al., 2003), despite their occurrence in different multiprotein complexes that exhibit alternative lysine target specificity.

Biochemical purifications from both *Drosophila* and human cells have demonstrated the distinct nature of complexes containing either ADA2a or ADA2b despite their relatively high level of sequence similarity (Barlev et al., 2003; Kusch et al., 2003; Guelman et al., 2006; Gamper et al., 2009). The ADA2b protein is exclusively found in the STAGA complex while its paralog, ADA2a, is excluded from these complexes but associates instead with the ATAC and PCAF complexes. Interestingly, the adaptor protein ADA3, which is a well-characterized component of the acetyltransferase catalytic core in complexes containing GCN5, is present in both ADA2a- and ADA2b-containing complexes. Deletion studies in *Drosophila* have demonstrated that the different ADA2 paralogs play a regulatory role in the targeting of GCN5-containing complexes to individual histone substrates in the context of chromatin (Qi et al., 2004). Deletion of Ada2a does not affect the acetylation of histone H3; however, it does compromise the acetylation of histone H4 at specific lysine residues. Interestingly, loss of Ada2b results in a significant reduction in the acetylation of lysine residues on histone H3 targeted by GCN5 but does not appear to alter the acetylation of histone H4 (Qi et al., 2004). Further, deletion of GCN5 also causes a loss in histone H4 acetylation, which overlaps with that seen in the ADA2a deletion mutant (Ciurciu et al., 2006). These results imply a role for the ADA2a-regulated, GCN5-mediated acetylation of histone H4 and further that the substrate specificity of the GCN5 acetyltransferase can be specifically modulated by its interaction with either of the ADA2 paralogs.

IV. SAGA and Related Complexes Possess Histone Deubiquitination Activity

The activity of Gcn5 in the context of the catalytic heterotrimer that it forms with Ada2 and Ada3 represents a critical function within the greater SAGA complex; however, SAGA-related complexes exhibit several distinct chromatin-modifying activities and therefore play multiple roles in the regulation of transcription. Several functions in addition to acetyltransferase activity are afforded to the SAGA as a result of the presence of the aforementioned modules, which contribute to the overall composition of the complex. One of these SAGA modules exhibits ubiquitin hydrolase activity and is involved in the regulation of histone H2B monoubiquitination (Köhler et al., 2006, 2008; Lee et al., 2009). Posttranslational modification of the histone H2B C-terminal tail by monoubiquitination is catalyzed by Rad6, a ubiquitin ligase, in yeast or its homologs in other organisms (Robzyk et al., 2000; reviewed in Osley, 2006). Rad6-mediated monoubiquitination of histone H2B is dependent on the recruitment of both RNA polymerase II (RNAPII) and other transcription factors and has been shown to influence di- and trimethylation levels of histone H3 (reviewed in Weake and Workman, 2008). The SAGA subunit Ubp8 and its homologs Nonstop and USP22 in *Drosophila* and humans, respectively, are ubiquitin hydrolases that can specifically remove the monoubiquitin moiety from histone H2B (Henry et al., 2003; Daniel et al., 2004; Weake et al., 2008; Zhao et al., 2008). In yeast, Ubp8 physically interacts with accessory proteins, Sgf11 and Sus1, to form the functional deubiquitination module (hereafter referred to as the Ubp8 module) that is found in association with the SAGA complex (Köhler et al., 2006, 2008; Lee et al., 2009). Similarly, USP22, the human homolog of Ubp8, is bound to the STAGA complex through interactions with ATXN7L3 and ENY2, which are homologs of Sgf11 and Sus1, respectively (Zhao et al., 2008). These histone H2B deubiquitination modules, Ubp8/Sgf11/Sus1 in yeast and USP22/ATXN7L3/ENY2 in humans, are anchored to SAGA complexes through their interactions with homologs of the ataxin-7 protein from their respective organisms.

While neither the deubiquitination of histone H2B nor the presence of the deubiquitination module itself appear to overtly influence the HAT activity of their parent SAGA complexes, the monoubiquitination of histone H2B plays an important role in regulating transcription. In fact, the presence of a functional STAGA deubiquitination module has been directly implicated in nuclear receptor-mediated transcriptional activation (Zhao et al., 2008). Deletion analysis in yeast has demonstrated that Ubp8 is required for the recruitment of Ctk1 kinase and its subsequent phosphorylation of the RNAPII C-terminal domain (CTD) on serine 2 within the heptapeptide-repeat domain at SAGA-regulated genes (Wyce et al., 2007). Phosphorylation of the RNAPII CTD is required for the initiation and elongation of transcription by RNAPII, making the function of Ubp8 critical for proper regulation of SAGA-dependent gene expression. Additionally, the studies by Wyce et al. (2007) indicated that the persistence of monoubiquitinated histone H2B, due to the deletion of Ubp8, prevented the recruitment of the Set2 methyltransferase and as a result led to a reduction in levels of histone H3 methylation at lysine 36, a mark associated with transcriptionally active genes (Shilatifard, 2006). Given the regulatory effect that monoubiquitinated histone H2B exerts on transcriptional initiation and subsequent elongation of the nascent transcript by RNAPII, it seems clear that the SAGA complexes facilitate transcriptional activation and elongation not only through acetylation of chromatin, but also through the modulation of histone ubiquitination.

V. Gcn5 and PCAF Acetyltransferases Exhibit Distinct, Yet Partially Overlapping, Functions

As discussed above, both Gcn5 and PCAF are highly homologous proteins that serve as acetyltransferases in different multiprotein complexes. Somewhat predictably due to their high degree of sequence similarity, Gcn5 and PCAF exhibit partially overlapping, yet distinct, functions within the cellular context. While the PCAF complex contains a similar complement of proteins as the Gcn5-containing complexes, there are several differences that may explain both their overlapping and dissimilar functions. To better understand the roles of both Gcn5 and PCAF, the effects of their deletion on the development and adult phenotype of transgenic, knock-out mice were examined. Embryogenesis is a highly regulated process during which the up- and downregulation of many genes is required for the proper formation of organs and tissues within the developing animal. While the functions of Gcn5 had been well studied in both yeast and cultured mammalian cells, the role of this acetyltransferase, and its closely related homolog PCAF, in developmental processes was not as well characterized. To address questions relating to the roles of these enzymes in mouse development, several transgenic models were generated harboring deletions of the Gcn5 locus, the PCAF locus, or both. Despite the high similarity between the two proteins, their control over embryogenesis was shown to be quite disparate. Surprisingly, homozygous deletion of the PCAF gene resulted in a lack of physical impairment in adult mice (Yamauchi et al., 2000). Loss of PCAF expression, however, did lead to neurological dysfunction in young adult mice, including memory deficit and an exaggerated response to externally applied stress (Maurice et al., 2008). In contrast, despite a lack of phenotype in heterozygous Gcn5 knock-out mice, loss of both Gcn5 alleles led to embryonic lethality that was associated with an increase in apoptosis and a failure to complete neural tube development (Xu et al., 2000; Yamauchi et al., 2000). Homozygous deletion of both PCAF and Gcn5 alleles led to embryonic lethality occurring at earlier stages in development than that observed in Gcn5-null mice, further suggesting that the two proteins were required for functionally distinct processes. Not surprisingly, considering the distinct phenotypes of Gcn5- and PCAF-null mice, Gcn5 mRNA was found to be highly expressed until later stages in embryonic development while PCAF mRNA was barely detectable until mice reached adulthood (Xu et al., 2000). Additional analysis of tissues from these mouse models indicated that expression of Gcn5 protein increased in response to loss of PCAF (Yamauchi et al., 2000). Coupled with the more severe developmental phenotype observed in mice lacking both PCAF and Gcn5 it can be concluded that these two acetyltransferases exhibit distinct but partially overlapping functions. These findings indicate a vital role for the Gcn5 protein, even at a reduced level of expression, in the completion of proper nervous system development and suggest two possible avenues for

Gcn5 dependence during mouse embryonic development. The first possibility is that the requirement for specific acetyltransferases at individual gene loci may be flexible in the case of certain genes and more stringent for others. Some genes may specifically require the activity of Gcn5 to be properly regulated, while others may be appropriately regulated by either PCAF or Gcn5. This scenario may explain why Gcn5 is upregulated in PCAF-null mice and how these animals can complete development normally. PCAF may be a more "dispensable" acetyltransferase during embryonic development since it has arisen more recently, evolutionarily speaking, compared to its paralog Gcn5, which is found in more ancestral organisms. A second possibility is that the Gcn5 protein may be providing or facilitating a yet-to-be discovered activity that is essential for the normal regulation of genes required for developmental processes.

The idea that Gcn5 and PCAF have overlapping functions has been further supported by experiments performed in cultured chicken cells, which demonstrated a robust upregulation of PCAF expression in response to loss of Gcn5 (Kikuchi et al., 2005). The observed compensation for loss of Gcn5 was PCAF-specific since the expression of other acetyltransferases was not significantly altered. It is not difficult to believe that loss of either Gcn5 or PCAF results in the increased expression of the other since these two proteins share a very high degree of sequence similarity and function in closely related, multiprotein complexes. Interestingly, the altered steady-state levels of either Gcn5 or PCAF observed after loss of one or the other apparently compensates for the overall level of acetyltransferase protein but leads to differential patterns of histone H3 and H4 acetylation (Kikuchi et al., 2005). While not unexpected, this observation does raise the question of how compensation of Gcn5, or PCAF, function is truly achieved in the absence of a known pattern of overlapping histone acetylation. The answer could suggest that the generalized acetylation of chromatin by these proteins may play a more important role in regulating gene expression than Gcn5- or PCAF-specific acetylation.

VI. Polyglutamine Stretches are Common Attributes of Transcription Factors

The initiation of gene transcription is an intricate process involving the coordinated actions of both protein–DNA and protein–protein interactions. DNA-binding transcriptional activators have been shown to minimally contain one of four general domains, acidic, glutamine-rich, proline-rich, and DNA binding, that influence their ability to promote transcription (Courey and Tijan, 1988). Of the aforementioned four domains, the acidic, glutamine-rich, and proline-rich domains have been shown to impact the ability of a protein to facilitate transcription (Mitchell and Tijan, 1989). The obvious necessity for the DNA-binding domain aside deletion of any of the three remaining domains negatively impacts the ability of a transcription factor to promote transcription (Courey and Tijan, 1988; Gerber et al., 1994). The significance of glutamine-rich domains in the transactivation potential of transcription factors was explored on the global level in the budding yeast *Saccharomyces cerevisiae* by two high-throughput screens that yielded partially complementary results (Uetz et al., 2000; Ito et al., 2001). The amino acid sequences of transcriptional activators that were identified in these screens were analyzed to identify patterns that correlated with their transactivation potential. While there were several physiochemical properties that appeared to be common to many of these proteins, those that contained a glutamine-rich region were shown to be strongly associated with a greater potential to stimulate transcription (Titz et al., 2006). More detailed studies in mammalian cells have shown that the presence of glutamine-rich domains in transcriptional activators, such as the cAMP response element binding (CREB) protein and Sp1, are necessary for their association with chromatin *in vivo* (Mayr et al., 2005). Additionally, glutamine-rich domains are interchangeable between these transcriptional activators, indicating their overall importance in this context.

Interrogation of information within protein databases for the identities and functions of proteins that contain homopolymeric tracts of glutamine (containing at least 20) or proline (containing at least 10) residues revealed that the majority of hits returned, 82% and 78%, respectively, were factors associated with transcription (Gerber et al., 1994). Exploration of these findings indicated that an increase in the number of glutamine or proline residues in their respective homopolymer tracts correlated with an increased ability to stimulate transcription *in vitro*. Interestingly, the effect of increasing homopolymeric glutamine tract length on transcriptional activation was limited to peptides containing fewer than 40 glutamines (Gerber et al., 1994). A group of otherwise unrelated genes have been shown to be subject to a mutation that results in lengthening of polyglutamine tracts within their protein products. Polyglutamine expansion of these proteins results in altered function and polyglutamine tract-dependent aggregation of proteins to form nuclear inclusions (reviewed in Orr and Zoghbi, 2007). Understanding the normal role of glutamine-rich domains within transcriptional activators is of particular interest in the context of elucidating the pathological mechanisms associated with polyglutamine-expansion disorders. One of these diseases, spinocerebellar ataxia type 7 (SCA7), will be the discussed below.

VII. Polyglutamine-Expansion Disorders

The polyglutamine-expansion disease family encompasses at least nine heritable disorders (Table I), including Huntingtin's disease (HD), dentatorubral pallidoluysian atrophy (DRPLA), spinal and bulbar muscular atrophy (SBMA), and the spinocerebellar ataxias SCA1, SCA2, SCA3, SCA6, SCA7, and SCA17 (reviewed in Orr and Zoghbi, 2007). Each of these disorders results from the expansion of a polyglutamine tract that is present in the wild-type protein. Polyglutamine expansion of disease-associated proteins is the direct effect of instability within the CAG trinucleotide

Table I
Basic Information on Genes that are Associated with Polyglutamine Expansion Disorders

Disease	Gene	Normal CAG repeat length	Pathogenic CAG repeat length	Normal protein function
HD	Huntingtin	6–34	36–121	Multifunctional scaffold
SCA1	Ataxin-1	8–44	39–83	Transcriptional regulation
SCA2	Ataxin-2	13–33	32–77	RNA metabolism
SCA3	Ataxin-3	12–40	54–89	Ubiquitin protease
SCA6	CACNA-1	4–18	19–33	Voltage-gated calcium channel
SCA7	Ataxin-7	4–35	37–306	Structural component of GCN5 HAT complexes
SCA17	TBP	29–42	47–55	General transcriptional factor
SBMA	Androgen receptor	6–39	40–63	Nuclear receptor
DRPLA	Atrophin-1	6–36	49–84	Unknown

coding region at the genomic locus of the affected gene. Proteins that are subject to polyglutamine expansion are thought to undergo conformational change (reviewed in Nagai and Popiel, 2008) and in some cases proteolytic cleavage that results in products exhibiting a cytotoxic function (Wellington et al., 1998; Ellerby et al., 1999). The autosomal dominant inheritance patterns observed in eight of the nine polyglutamine-expansion disorders suggest that the altered proteins associated with disease exert a functional dominance over their wild-type counterparts. Interestingly, polyglutamine-expanded proteins are widely expressed throughout the body; however, pathology is primarily restricted to neuronal tissue. The mechanisms of pathology are not precisely known for members of this disease family, although many of the genes affected encode proteins that are normally involved in transcriptional regulation in some capacity. The following will discuss the implications of findings from the literature on the elucidation of mechanisms associated with the pathology of polyglutamine-expansion disorders. An emphasis will be placed on SCA7 since the protein that becomes expanded in this condition, ataxin-7 (ATXN7), is an integral component of the SAGA family of acetyltransferase complexes that are conserved from yeast to humans.

VIII. Spinocerebellar Ataxia Type 7

Spinocerebellar ataxia type 7, formerly known as autosomal-dominant cerebellar ataxia with retinal degeneration (ADCA Type II), is in the family of neurodegenerative disorders known as trinucleotide repeat, or polyglutamine-expansion, disorders. The gene associated with SCA7 was mapped to chromosome 3p12-p13 (Benomar et al., 1995; David et al., 1996) and further examination of the genomic region indicated that the disease-associated allele was subject to expansion of a CAG trinucleotide repeat sequence (Lindblad et al., 1996). Cloning of the gene responsible for SCA7, hereafter referred to as ataxin-7 (ATXN7), followed and demonstrated a broad range of CAG repeats, between 38 and 130, associated with development of SCA7 within individuals that were surveyed in the study (David et al., 1997). Trends observed in families harboring the pathogenic ATXN7 allele demonstrated that the age of symptom onset decreased, which was accompanied by an increased rate of disease progression, in affected offspring compared to their parents, suggesting clinical anticipation. Additionally, paternal transmission of the mutant gene resulted in greater anticipation and therefore an earlier age of onset when compared to maternal transmission, which raises the possibility of paternal imprinting at the ATXN7 gene locus (Benomar et al., 1994).

Trinucleotide repeat expansion within the ATXN7 gene results in a protein product containing a polyglutamine-expanded region. Clinically, polyglutamine expansion of the ATXN7 protein is characterized by progressive degeneration of both the cerebellum and the retina. Cerebellar degeneration leads to ataxia, which is characterized by a generalized reduction in the ability to coordinate muscle movements. Less-penetrant phenotypes include muscle weakness, amyotrophy, difficulty speaking (dysarthria), difficulty swallowing (dysphagia), hearing loss, and intellectual impairment (Benomar et al., 1994). While ataxia is a symptom that is common to all of the trinucleotide repeat expansion disorders, SCA7 is unique in the additional development of retinal degeneration, which is characterized by cone-rod dystrophy and eventual blindness. The highly complex nature of cellular dysfunction in the context of SCA7 has led to the development and use of many experimental systems and the publication of studies that contribute to the understanding of disease mechanisms from multiple experimental approaches.

IX. Transgenic Mouse Models of SCA7

Transgenic, knock-in mouse models of SCA7 have been produced with expression of polyglutamine-expanded ATXN7 in Purkinje cells (Yvert et al., 2000), retina (Yvert et al., 2000; La Spada et al., 2001), whole brain (Yvert et al., 2001), central nervous system with the exception of Purkinje cells (La Spada et al., 2001), Bergmann glia of the cerebellum (Custer et al., 2006), or throughout the entire body (Yoo et al., 2003). The transgenic mouse models that have been designed to express mutant ATXN7 in different regions of the brain exhibit several commonalities that mimic clinical manifestations of SCA7 in humans. With the exception of those expressing polyglutamine-expanded ATXN7 under control of a retinal-specific promoter, these mice demonstrated symptoms of progressively worsening ataxia. Further, the age of onset and rate of degeneration varied depending on the specific tissue in which the polyglutamine-expanded ATXN7 protein was expressed. In addition to variability relating to the tissue in which the mutant ATXN7 protein was expressed, the observed disparity in disease progression in these transgenic mouse models was likely to have resulted from the different polyglutamine tract expansion lengths used in each model.

Purkinje cell-specific expression of mutant ATXN7 led to the observation of a detectable phenotype in transgenic mice at approximately 11 months of age (Yvert et al., 2000). Despite the observation of impaired motor function in these animals, histological examination of cerebellum tissue samples from 11-month-old individuals did not show obvious signs of degeneration in Purkinje cells, or any other cell type within the cerebellum. In this model, nuclear accumulation of polyglutamine-expanded ATXN7 protein was observed when mice were 1 month of age, followed by the formation of nuclear inclusions in rostral lobes of the cerebellum at 4 months. By the time mice were 16 months of age, most Purkinje cells contained at least one nuclear inclusion. Inclusions observed at later time points appeared to localize adjacent to the nucleolus when observed by either indirect immunofluorescence or electron microscopy. In contrast to observations made in their 11-month-old counterparts, Purkinje cells in cerebellar tissue sections obtained from 16-month-old mice exhibited a reduction in dendritic arborization and were present in a generally lower density compared to matched controls (Yvert et al., 2000). Expression of polyglutamine-expanded ATXN7 in Bergmann glial cells, which are responsible for clearing glutamate from Purkinje cell synapses, also led to the development of ataxia in transgenic mice (Custer et al., 2006). Bergmann glia expressing mutant ATXN7 exhibited a reduction in levels of the glutamate transporter GLAST and glutamate uptake. Purkinje cells associated with Bergmann glia expressing polyglutamine-expanded ATXN7 demonstrated signs of dark cell degeneration, which is associated with excitotoxicity linked to glutamate overstimulation (Leranth and Hamori, 1970; Strahlendorf et al., 1998; Strahlendorf and Strahlendorf, 1999). These two studies show that neurological dysfunction associated with expression of polyglutamine-expanded ATXN7 can result from either cell autonomous or non-cell-autonomous Purkinje cell degeneration.

Transgenic mice expressing mutant ATXN7 in the central nervous system with the exception of Purkinje cells exhibited symptoms of neurological dysfunction, including ataxia, beginning as early as 8 weeks of age (Garden et al., 2002). Sections of cerebellum prepared from

20-week-old individuals demonstrated a reduction in the size and dendritic arborization of Purkinje cells, despite lack of mutant ATXN7 protein expression. Further, apoptotic cells were not observed in these tissue sections, indicating the neurological dysfunction in these individuals occurred in the absence of significant neuronal cell loss. The onset of neurological symptoms in these transgenic mice correlates with the appearance of nuclear inclusions in the granule and molecular cell layers of the cerebellum, along with a reduction in the detection of the Purkinje cell marker calbindin by immunohistochemistry (Garden et al., 2002). The results of this study are of particular importance because they demonstrate that Purkinje cell degeneration in SCA7 can occur in a non-cell-autonomous manner and therefore may indicate a more significant role of mutant ATXN7 expressed in other cell types of the cerebellum in SCA7 pathology. Consistent with these findings, Bergmann glia-specific expression of mutant ATXN7 was sufficient to produce ataxia and degeneration (Custer et al., 2006), suggesting that glial cell dysfunction leads to Purkinje cell degeneration.

Transgenic mice expressing polyglutamine-expanded ATXN7 in all tissues of the brain demonstrated yet another timeline for disease onset and progression (Yvert et al., 2001). Individuals with whole-brain expression of mutant ATXN7 were observed to exhibit initial symptoms of neurological dysfunction between 3 and 8 months of age. The formation of nuclear inclusions in these animals often coincided, or preceded, the onset of disease symptoms and progressed in a manner similar to that observed in other transgenic mouse models. Additionally, the distribution of nuclear inclusions between affected and unaffected neuronal populations was indistinguishable, suggesting that the formation of nuclear inclusions was not directly linked to toxicity in all cell types (Yvert et al., 2001).

The pathology of SCA7 is set apart from the other polyglutamine-expansion disorders by the occurrence of progressive retinal degeneration in affected patients. To better understand the mechanisms of SCA7 retinopathy in the absence of other neurodegenerative phenotypes, a transgenic mouse model was developed in which either normal or polyglutamine-expanded ATNX7 was expressed under control of the retinal-specific rhodopsin promoter (Yvert et al., 2000). Following initial characterization of the retinal-specific ATXN7 transgenic model, subsequent studies using the same transgenic mouse line demonstrated that these mice exhibited loss of ATXN7 transgene expression during the first few months of their lives. The reduction and eventual loss of transgenic ATXN7 expression was shown to result from downregulation of the rhodopsin promoter by the mutant ATXN7 protein (Helmlinger et al., 2004a). These observations led to the identification of this model as a system for studying how loss of polyglutamine-expanded ATXN7 at an early age influenced the onset and progression of SCA7 retinopathy. Rhodopsin promoter activity, affecting both endogenous rhodopsin and ATXN7 transgene expression, was greatly reduced in 3-week-old mice and continued to decline with age until becoming negligible by the time mice reached 9 weeks of age. Despite the loss of detectable exogenous ATXN7 mRNA and soluble ATXN7 protein by week 9, nearly all of the cells in retinal tissue from 10-week-old mice contained nuclear inclusions that stained positively with antibodies specific for the N-terminal region of the ATXN7 protein. Nuclear inclusions were also observed in these mice up to 2 years of age, indicating their maintenance over time despite early loss of mutant ATXN7 expression. The persistence of nuclear inclusions in this SCA7 mouse model is of particular interest because the breakdown of nuclear inclusions in a mouse model of HD was shown to occur in as few as 3 weeks after loss of polyglutamine-expanded Huntingtin protein expression (Martin-Aparicio et al., 2001). The lack of agreement between the nuclear inclusion dynamics observed in these two distinct mouse models suggests that there is a fundamental difference in the way that cells handle these highly complex structures when they are initiated by different proteins. Interestingly, photoreceptor cell loss did not coincide with the downregulation of rhodopsin promoter activity in young mice; however, retinal dysfunction did begin several weeks later and progressed to a complete loss of photoreceptor function despite the eventual abolition of exogenous ATXN7 expression. These results indicate that the presence of ATXN7 nuclear inclusions in photoreceptors of the retina, but not continuous expression of the expanded ATXN7 gene, is required for both repression of the rhodopsin promoter and retinal dysfunction.

Cone-rod dystrophy and subsequent retinal degeneration are hallmarks of SCA7 disease that are observed in both human and transgenic mouse models. Outside the context of SCA7, cone-rod dystrophy has been clinically associated with the presence of mutations in the cone-rod homeobox (CRX) transcription factor, which is involved in the regulation of retinal-specific genes (reviewed in Hennig et al., 2008). La Spada et al. (2001) produced a transgenic, knock-in mouse model with the ATXN7 transgene under control of the neuron-specific murine prion protein promoter to study the phenotype that develops when polyglutamine-expanded ATXN7 expression is limited to neuronal tissue. Similar to other SCA7 mouse models, mice expressing polyglutamine-expanded ATXN7 in this model developed progressive retinal degeneration, which was accompanied by photoreceptor cell loss. While both normal and mutant ATXN7 were shown to interact with CRX, polyglutamine-expanded ATXN7 recruited CRX into nuclear inclusions. Sequestration of CRX by mutant ATXN7 was also coincident with reduced expression of several retinal-specific genes. Repression of these retinal-specific genes was attributed to a mutant ATXN7-dependent inhibition of the interaction between CRX and the rhodopsin promoter, which was observed in electrophoretic mobility shift assays. Subsequent studies further dissected the physical interaction between ATXN7 and CRX and its effect on the expression of retinal-specific genes.

Like many transcriptional activators, the CRX protein contains a polyglutamine-rich region; however, it has been shown to be dispensable for CRX-dependent activation of retinal-specific promoters (Chen et al., 2004). Expectedly, the interaction between CRX and ATXN7 is mediated through the glutamine-rich region of CRX and the polyglutamine domain in the N-terminus of ATXN7. Despite the requirement of the ATXN7 N-terminal region for the ATXN7–CRX interaction, the C-terminus of the ATXN7 protein is essential for maximal repression of CRX-dependent promoters by mutant ATXN7 (Chen et al., 2004). The necessity of the ATXN7 C-terminal region in the modulation of CRX-dependent promoters may reflect either its role in anchoring ATXN7 to the SAGA complex (Köhler et al., 2008) or the presence of nuclear localization signals within this region of the protein (Chen et al., 2004).

The ATXN7 protein was later shown to be required for the interaction of CRX with the STAGA complex and its subsequent recruitment to CRX-dependent promoters (Palhan et al., 2005). Additional experiments conducted on the retinas of transgenic mice by Palhan et al. (2005) demonstrated that the expression of polyglutamine-expanded ATXN7 led to reduction in the recruitment of CRX to some, but not all, retinal-specific promoters. These promoters were shown to be hypoacetylated on histone H3 on lysine 9 and 14; however, the reduction in histone H3 acetylation was often more dramatic than the corresponding change in CRX occupancy at the same promoter. To elucidate this apparent discrepancy, the presence of GCN5 at the promoters of interest was determined. In contrast to the reduced recruitment of CRX and acetylation of histone H3, the occupancy of GCN5 at the regulatory regions of these genes was not significantly different when either normal or mutant ATXN7 was being expressed. This finding indicated that the observed hypoacetylation was not the result of a defect in the ability of mutant ATXN7-containing STAGA complexes to effectively associate with target genes (Palhan et al., 2005). From these findings the authors concluded that polyglutamine-expanded ATXN7 was inhibiting the acetyltransferase activity of STAGA complexes without affecting their recruitment to chromatin and association with tissue-specific transcription factors. In addition to the loss of normal histone acetylation by STAGA, the presence of these nonfunctional complexes may prevent compensatory acetylation of these regions by other redundant or overlapping acetyltransferases implying a dominant-negative effect of mutant ATXN7. Hypoacetylation of histone H3 is associated with a lack of transcriptional activation and provides a mechanistic explanation of how polyglutamine-expanded ATXN7 expression leads to aberrant regulation of retinal-specific genes in both affected humans and mouse models. These results also highlight the importance of HAT complex recruitment and function in the activation of tissue-specific promoters and how their dysregulation can result in disease onset and progression.

X. Proteolytic Processing of ataxin-7

The formation and accumulation of ATXN7-containing nuclear inclusions over time was observed in all transgenic mouse models. Nuclear inclusions from several of these mouse models were shown to immunostain positive with antibodies recognizing either the N-terminal region of the protein or the polyglutamine tract, but not the ATXN7 C-terminal region (Yvert et al., 2000, 2001; Garden et al., 2002). These observations suggest that the N-terminal region of the ATXN7 protein, which contains the polyglutamine tract, is enriched in these protein aggregates. Later studies demonstrated that like several other polyglutamine-expanded proteins, ATXN7 is subject to cleavage by activated caspase proteases. These experiments established that both normal and polyglutamine-expanded ATXN7 are substrates of activated caspase-7 and cleavage can occur at either of two consensus sites within the ATXN7 protein. Other proteins that can undergo polyglutamine expansion, such as the androgen receptor (AR), atrophin-1, huntingtin, and ataxin-3, are also cleaved in a caspase-dependent manner, suggesting that this event may represent an important step in the cellular pathology associated with their respective diseases (Wellington et al., 1998). In some cases, such as that of the polyglutamine-expanded AR and ATXN7, caspase-mediated cleavage of polyglutamine proteins is required for their formation of nuclear inclusions (Ellerby et al., 1999; Young et al., 2007).

Caspase-7-mediated proteolysis of ATXN7 is required for polyglutamine-induced aggregation, transcriptional dysregulation, and cytotoxicity (Young et al., 2007). Cleavage of the ATXN7 protein at one of two known caspase-7 consensus motifs examined by Young et al. (2007) yields a product that corresponds to the N-terminal ATXN7 fragment that was observed by Garden et al. (2002) in extracts prepared from transgenic mice expressing mutant ATXN7. Interestingly, the preferred site of caspase-7-mediated cleavage of ATXN7 *in vivo* differs from the preferred site *in vitro*. The differential preference of caspase-7 for the consensus site that is preferred *in vivo* may reflect the influence of accessory factors that were absent during *in vitro* experiments. The requirement for proteolytic cleavage of ATXN7 prior to its inclusion into nuclear aggregates may indicate that the N- and C-terminal regions of the protein serve different functions in the cell or that protein turnover is regulated by proteolytic processing. Either way, the processing of the ATXN7 protein adds an additional level of complexity to the overall understanding of how it functions in both normal and disease states.

XI. Ataxin-7 is a Subunit in SAGA and SAGA-Related Histone Acetyltransferase Complexes

The function of the normal ATXN7 protein has yet to be clearly determined; however, it has been shown to reside in both yeast and human HAT complexes and has been implicated as a potential transcriptional activator by global proteomics studies performed in yeast (Titz et al.,

2006). Human ataxin-7 (hATXN7) and its yeast homolog, Sgf73, hereafter referred to as yATXN7, are constituents of the SAGA/SLIK and STAGA/TFTC HAT complexes, respectively (Sanders et al., 2002; Helmlinger et al., 2004b; McMahon et al., 2005; Palhan et al., 2005). Normal fractionation of SAGA is disrupted by deletion of yATXN7 despite maintenance of HAT activity on free core histones by recoverable complexes. Interestingly, deletion of yATXN7 abolishes the ability of purified SAGA to acetylate histones in the context of the nucleosome octamer *in vitro*. Restoration of nucleosomal HAT activity is observed in chromatographically purified samples when deletion of yATXN7 is complemented by expression of normal human ATXN7 (hATXN7), but not by polyglutamine-expanded hATXN7 (McMahon et al., 2005). These findings were mirrored when transiently expressed normal and polyglutamine-expanded ATXN7 were purified and assayed for their relative abilities to acetylate oligonucleosomal substrates *in vitro* from cultured human cells (Palhan et al., 2005). Maintenance of free HAT activity accompanied by loss of the ability to acetylate nucleosomes suggests that polyglutamine expansion of the ATXN7 protein in either species disrupts the substrate targeting of SAGA complexes, but not acetyltransferase activity of GCN5. To lend additional support for the role of ATXN7 in targeting the SAGA complex to its nucleosome substrate, yATXN7 has been shown to be essential for the recruitment of SAGA to the upstream activation sequences of genes that it regulates. Further, yATXN7 is required for the recruitment and assembly of the transcriptional preinitiation complex (PIC) at the promoters of SAGA-dependent genes in a HAT-dependent or independent manner (Shukla et al., 2006).

Normal and mutant ATXN7 have been shown to copurify with the Gcn5 protein in yeast and both GCN5 isoforms in human cells (McMahon et al., 2005; Palhan et al., 2005). As discussed above, the longer GCN5 isoform is thought to represent a differentially spliced GCN5 transcript and has been shown to contain an additional N-terminal domain that is highly homologous to the N-terminal domain of its homolog PCAF. The additional N-terminal domain allows GCN5-L to recognize and acetylate histones in the context of nucleosomes, which is a capability that is not exhibited by its shorter homolog when assayed *in vitro* with recombinant protein (Xu et al., 1998). Given the copurification of hATXN7 and GCN5-L and the appearance of GCN5-L as the preferred GCN5 isoform in STAGA, it seems likely that mutant hATXN7 could disrupt the normal acetylation of chromatin by interfering with the nucleosomal targeting ability of GCN5-L or the STAGA complex as a whole. Alternatively, improper substrate recognition may result from the depletion of specific SAGA subunits, including Ada2/Ada2b, Ada3, TAF12, and Spt3 in both yeast and humans (McMahon et al., 2005; Palhan et al., 2005). Loss of these subunits may be of particular importance to the mutant yATXN7-dependent SAGA dysfunction in yeast due to the absence of the N-terminal domain and subsequent inability of yGcn5 to target nucleosomal substrates without the aid of other factors. Consistent with this observation, Ada3 and TAF12 are required for Gcn5 to efficiently acetylate histones within nucleosomes (Grant et al., 1998; Balasubramanian et al., 2002). Of additional importance, STAGA complexes that copurify with polyglutamine-expanded hATXN7 are deficient in the adaptor protein ADA2b (Palhan et al., 2005). Exclusion of ADA2b from the mutant STAGA complex would also be expected to significantly alter the targeting and substrate specificity of its GCN5-dependent acetyltransferase activity. This could be expected to result in either generalized loss of GCN5-dependent histone acetylation or hyperacetylation of aberrant histone substrates. Exclusion of ADA2b also has the potential to alter chromatin occupancy of mutant ATXN7-containing STAGA. The dominant-negative function of polyglutamine-expanded ATXN7, suggested by Palhan et al. (2005), may reflect a situation in which the mutant STAGA complex binds to chromatin, is unable to acetylate associated nucleosomes, and fails to dissociate, therefore impeding other transcription factors from accessing chromatin efficiently. Alternatively, loss of ADA2b may selectively prohibit mutant STAGA from interacting with its chromatin substrate at certain targets and, as a result, facilitate the aberrant acetylation of normal STAGA targets by alternative, promiscuously acting acetyltransferase complexes. Abolition of the normal enzyme–substrate interactions between STAGA and nucleosomal histones is likely to facilitate the establishment of atypical chromatin acetylation patterns from either similar, yet nonredundant, acetyltransferases, such as PCAF, or other more dissimilar acetyltransferases.

Several studies have demonstrated the loss of nucleosomal HAT activity in polyglutamine-expanded ATXN7-containing SAGA complexes (McMahon et al., 2005; Palhan et al., 2005). Interestingly, Helmlinger et al. (2006) observed severe decondensation of chromatin within the nuclei of retinal cells expressing polyglutamine-expanded ATXN7, containing 90 glutamine repeats, under the control of a retinal-specific promoter. This observation was counterintuitive to the previously established loss of acetyltransferase activity due to the incorporation of mutant ATXN7 since chromatin decondensation is characteristic of an acetylated, transcriptionally active state. Follow-up experiments demonstrated that both STAGA/TCTF complex recruitment, as measured by chromatin immunoprecipitation using antibodies that recognize the STAGA subunit SPT3, and acetylation of histone H3 on lysine 9 and 14 increased significantly at retinal-specific genes that were dysregulated. Taken together, these observations indicate that STAGA/TFTC complexes containing polyglutamine-expanded ATXN7 would exhibit prolonged chromatin occupancy and lead to hyperacetylation of associated target genes. The inference of STAGA/TFTC presence as a result of SPT3 localization was accurate given the known association of SPT3 with these complexes at the time (Martinez et al., 1998; Martinez et al., 2001); however, more recently published data suggest the incorporation of SPT3 into other multiprotein complexes (Wang et al., 2008). While the aforementioned observations do not directly evaluate the ability of other SPT3-containing complexes to acetylate histones, they do demonstrate its coimmunoprecipitation with PCAF, in addition to GCN5, and distribution over a broad molecular weight range during size exclusion chromatography of human nuclear extracts. Of related interest, when yeast harboring a deletion of the yATXN7 gene were analyzed for the transcriptional activity of SAGA-dependent genes it was shown that the expression of yATXN7 protein was required for both the recruitment of intact SAGA to the UAS and the assembly of the transcriptional PIC (Shukla et al., 2006). In the same study, it was shown that deletion of yATXN7 resulted in an increased presence of lysine 9 and 14 acetylated histone H3 at SAGA-dependent promoters; however, the increased acetylation level was determined to be an artifact of reduced histone H3 eviction from these regions in the yATXN7-deleted strain. Later studies also suggest that Spt3 functions in a SAGA-independent manner to stabilize global levels of histones H3 and H4 in yeast strains harboring temperature-sensitive alleles of Not1, a structural component of the Ccr4–Not complex (James et al., 2007). In this scenario, the existence of a SAGA-independent role for Spt3, proposed by James et al. (2007), is also supported by their observation of a synthetic phenotype between *spt3*Δ and *spt7*Δ yeast strains since deletion of Spt7 is sufficient to abolish SAGA complex activity (Grant et al., 1997).

Tissues derived from SCA7 transgenic mice have been shown to exhibit a progressively decreasing amount of full-length, soluble polyglutamine-expanded ATXN7 protein, which is likely to reflect the results of both proteolytic processing by caspase-7 and incorporation of the mutant protein into insoluble inclusions over time. Interestingly, despite its preferential incorporation into nuclear inclusions, the abundance of a soluble N-terminal ATXN7 fragment appears to be comparable to its counterpart that is derived from nonexpanded ATXN7 (Garden et al., 2002). The significance of the N-terminal region of ATXN7 has yet to be fully characterized; however, it has been implicated in several important aspects of normal ATXN7 function, including physical interaction with both transcription factors (Chen et al., 2004) and the deubiquitination module of SAGA complexes (Köhler et al., 2008; Zhao et al., 2008). The net outcome is an apparent reduction in the availability of the soluble ATXN7 C-terminal region, which is required for the interaction between ATXN7 and the SAGA complex. Therefore, decreased availability of the ATXN7 C-terminus may have a substantial effect on the regulation of SAGA-dependent genes since it is required not only for the structural integrity of the SAGA complex, but also for its ability to efficiently associate with its chromatin substrate. Further, recombinant, full-length polyglutamine-expanded ATXN7 was able to inhibit nucleosomal acetyltransferase activity of purified PCAF complex when added to *in vitro* acetylation reactions. Interestingly, addition of a polyglutamine-expanded N-terminal ATXN7 fragment had no effect under the same conditions (Palhan et al., 2005). These findings support the proposed interaction of ATXN7 with the SAGA complex through its C-terminal region and the importance of the C-terminal region in the ability of ATXN7 to control the activity of other acetyltransferase complexes. Regardless of the acetyltransferase, there is clear potential for disruption of the normal pattern of histone acetylation, in addition to other histone modifications that constitute the regulatory histone code by polyglutamine-expanded ATXN7. Despite its ability to interfere with nucleosomal acetylation, mutant ATXN7 does not directly abolish the ability of the GCN5 protein to catalyze the acetyltransferase reaction, since purified mutant yeast and human SAGA complexes retain the ability to acetylate free histones *in vitro* (McMahon et al., 2005; Palhan et al., 2005). Rather, incorporation of mutant ATXN7 into these complexes appears to disrupt their normal ability to acetylate histones within the context of nucleosomes, implying that polyglutamine expansion of ATXN7 disturbs the targeting or substrate presentation mechanisms of these complexes that allow them to interact with their physiologically relevant targets.

In yeast, the N-terminal region of yATXN7 is not required for its association with SAGA; however, it is required for its ability to link the Ubp8 deubiquitination module with the complex (Köhler et al., 2008; Lee et al., 2009). Additionally, purification of the N-terminal region of yATXN7 results in the purification of the Ubp8 deubiquitination module, but not the SAGA complex as a whole. Deletion studies in yeast demonstrated that the N-terminal region of yATXN7 is both necessary and sufficient for the association of the deubiquitination module, consisting of Sus1, Sgf11, and the deubiquitinating enzyme Ubp8, with other subunits of the SAGA complex. These findings in yeast are corroborated by similar studies implicating the necessity of ATXN7 for the association and activity of the deubiquitination module in STAGA (Zhao et al., 2008). As discussed above, these SAGA deubiquitination modules are involved in the removal of monoubiquitin from histone H2B, which is correlated with RNAPII elongation. It is also possible that the STAGA complex regulates transcription by modulating the ubiquitination states of other transcription factors in a manner similar to that of other polyglutamine-expansion proteins, such as ataxin-3 (ATXN3). The non-expanded ATXN3 protein functions to repress gene activation through the inhibition of CBP/p300- and PCAF-dependent histone acetylation and the recruitment and maintenance of transcriptional corepressors to chromatin (Li et al., 2002). Transcriptional repression mediated by ATXN3 is dependent on its ability to inhibit the proteosome-mediated degradation of corepressor complexes by both deubiquitylating polyubiquitylated constituent proteins and preventing recognition of these polyubiquitin chains by the ubiquitin-proteosome pathway (Berke et al., 2005; Burnett and Pittman, 2005). Polyglutamine expansion of ATXN3 does not significantly impact its interactions with the transcriptional repressors HDAC3 and NCoR *in vivo* or *in vitro*; however, it does inhibit the deacetylase activity of ATXN3-containing complexes, which is dependent on the ubiquitin protease activity of the ATXN3 protein (Evert et al., 2006).

While current evidence indicates that SAGA complexes are involved in the removal of monoubiquitin from histone H2B, the effect of polyglutamine-expanded ATXN7 on the integrity and association of the deubiquitination module with the SAGA complex has not yet been studied. The normal ATXN7 protein may also facilitate the targeting of other, nonhistone substrates for deubiquitination, or simply anchor the deubiquitination module to the greater complex. Either way, the potential for alteration of this SAGA module in molecular pathogenesis associated with polyglutamine expansion of ATXN7 is obvious and warrants additional exploration.

The multifunctional nature of the ATXN7 protein, and the SAGA complexes in which it resides, makes elucidation of its precise role in the molecular mechanisms of SCA7 pathology difficult. Clear associations have been made between the expression of polyglutamine-expanded ATXN7 protein and alteration of acetyltransferase activity in histone-modifying multiprotein complexes; however there appears to be other avenues by which this protein operates in the control of cellular functions. Interestingly, despite the structural alterations that accompany polyglutamine expansion of ATXN7, its association with transcription factors, such as CRX, does not appear to be significantly altered. In contrast, interaction with mutant ATXN7 prevents transcriptional activation by CRX and results in an altered gene expression profile that is associated with SCA7 retinal pathology. Although interactions with CRX are preserved, the inability of mutant ATXN7-containing protein complexes to effectively activate transcription is likely to be the result of lost associations with other proteins and therefore the inability to perform its normal functions. An example would be the disassociation of Ada2, Ada3, Spt3, and Taf12 from SAGA complexes containing mutant ATXN7 (Fig. 1). Loss of these subunits, particularly Ada2 and Ada3, would be likely to prevent the recognition and acetylation of nucleosomal substrates by Gcn5 without affecting the ability of the SAGA complex to associate with other factors. Interestingly, the scientific literature does not contain documented studies of the implications of ATXN7 polyglutamine expansion on the function of the SAGA complex deubiquitination module. It is easy to conceive that disruption of SAGA complex HAT activity would be sufficient to cause dysregulation of a large number of genes on its own; however, it is also difficult to ignore the potential for the additional alteration of deubiquitination activity given the data demonstrating the role of ATXN7 in anchoring the functional deubiquitination activity to the greater SAGA complex. Hopefully, future studies will elucidate whether polyglutamine expansion of ATXN7 alters the deubiquitination activity of the SAGA complex and its effects on the regulation

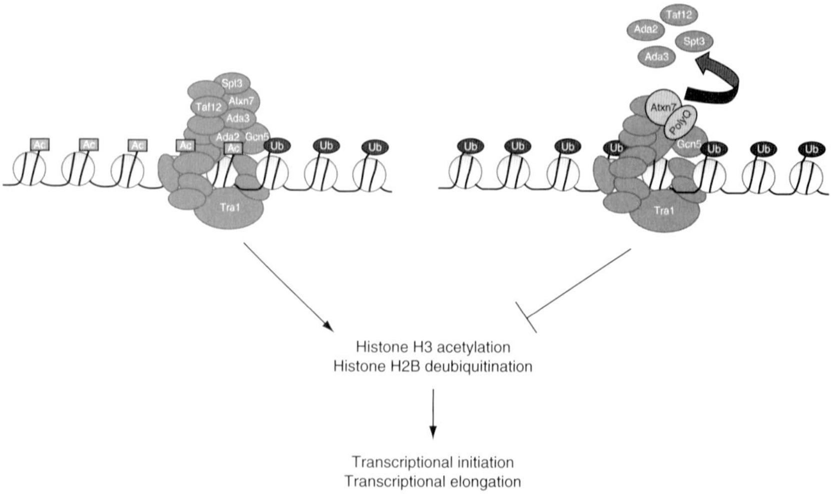

FIG. 1. Incorporation of polyglutamine-expanded ATXN7 into SAGA complexes alters subunit composition and catalytic activities. SAGA complexes containing polyglutamine-expanded ATXN7 exhibit reduced levels of Ada2, Ada3, Spt3, and TAF12. Several of these subunits, such as Ada2 and Ada3, are necessary for the ability of Gcn5 to acetylate nucleosomal histones. While experiments have not yet addressed the issue, there is potential for polyglutamine expansion of ATXN7 to disrupt the histone deubiquitination activity of SAGA complexes since the N-terminal region of ATXN7 is involved in the attachment of the deubiquitination module to the greater complex. Both histone acetylation and deubiquitination are involved in the initiation of gene transcription and subsequent elongation by RNA polymerase II.

of SAGA-dependent genes. The establishment of a better understanding of the molecular mechanisms underlying SCA7 pathology will facilitate the development of effective treatment methods for SCA7 and other polyglutamine-expansion disorders.

XII. Chemotherapeutic Options for the Treatment of Polyglutamine-Expansion Disorders

As highlighted above, substantial evidence exists to implicate the loss of HAT activity as a significant contributor to the pathology associated with polyglutamine-expansion disorders. As a result, the class of drugs known as histone deacetylase (HDAC) inhibitors has been tested as a therapeutic approach to the treatment of these disorders in laboratory models. HDAC inhibitors have been under investigation and development as chemotherapeutic approaches to the treatment of a variety of cancers with variable efficacy. Treatment with these drugs has been shown to reduce the expression of proteins involved in DNA synthesis, upregulate expression of proapoptotic factors, promote cell cycle arrest and differentiation, and more generally, modulate the expression of a broad range of genes (reviewed in Smith and Workman, 2009). While promising results have been observed, the specific mechanisms through which HDAC inhibitors influence cancer cell growth and overall disease progression are not yet fully clear. Fortunately, the development of these inhibitors as anticancer therapies has led to findings that support their use in the treatment of other diseases.

Sufficient evidence has been set forth in the scientific literature to support a link between altered histone acetylation and altered gene expression in several polyglutamine-expansion disorders. In this context, the application of HDAC inhibitors as potential chemotherapeutic agents for their treatment seems intuitive and is currently under exploration. Treatment of transgenic mice expressing different polyglutamine-expanded alleles, androgen receptor or atrophin-1, with the HDAC inhibitor sodium butyrate (SB) before the onset of symptoms demonstrated its ability to slow the rate of onset and overall disease severity (Minamiyama et al., 2004; Ying et al., 2006). In these experiments, brain tissue from transgenic mice was shown to exhibit an overall reduction in histone acetylation, especially of histone H3, compared to control mice that did not carry the polyglutamine-expanded allele. Treatment of these mice with SB resulted in an increase in histone H3 acetylation that was equivalent to or greater than matched wild-type controls, suggesting that SB treatment could offset the loss of histone acetylation that resulted from expression of these individual polyglutamine-expanded alleles. Further, the neuropathological events associated with onset of polyglutamine disease, such as neuronal loss and tissue degeneration, were either reduced or absent in drug-treated animals. While interpreting the results of HDAC inhibitor studies in mouse models of polyglutamine disease it is important to consider the effects of drug dosage on the outcome of treated animals. Interestingly, treatment of animals with higher, but still reasonable, doses of SB caused an earlier onset and more rapid progression of disease symptoms

(Minamiyama et al., 2004). Treatment of matched control animals with the same dose yielded no observable phenotype in the study. These observations suggest potential for the use of HDAC inhibitors in the treatment of polyglutamine-expansion disorders; however, they also emphasize the care that must be taken to establish effective dosing guidelines to prevent exacerbation of disease symptoms.

In addition to studies in transgenic animals, the effect of HDAC inhibitors in preventing, or reversing, cytopathology is being examined in cultured cells expressing polyglutamine-expanded proteins. Expression of polyglutamine-expanded ATXN7 in cultured cells leads to an increase in the number of cells undergoing apoptosis (McCampbell et al., 2001). Treatment of these cells with micromolar concentrations of SB resulted in a nearly 50% reduction in the number of apoptotic cells indicating a protective effect of this HDAC inhibitor on cells expressing the toxic polyglutamine-expanded ATXN7 protein. Similar results were obtained from experiments in the same study that were designed to test the efficacy of several other HDAC inhibitors, trichostatin A (TSA) and suberoylanilide hydroxamic acid (SAHA), in a cell culture model of SMBA. Surprisingly, treatment of cells in these experiments with yet another HDAC inhibitor, phenylbutyric acid (PBA), failed to prevent toxicity for reasons that were not determined. While still promising, studies examining the application of HDAC inhibitors in the treatment of HD have been less forthcoming than other cell culture models of polyglutamine-expansion disorders. Expression of a polyglutamine-expanded form of the Htt protein in cultured cells led to a decreased association of acetylated histone H3 with genes whose mRNA is specifically downregulated in HD models. These findings indicate a correlation between the hypoacetylation of histones and the repression of certain HD-associated genes. Interestingly, treatment of these cells with SB resulted in a global increase in acetyl histone H3, an increase in the association of acetylated histone H3 with affected genes, and the restoration of their normal expression levels (Sadri-Vakili et al., 2007). Further, the polyglutamine-expanded Htt protein has been shown to sequester the HAT protein CBP into aggregates in transgenic mouse models of HD (Steffan et al., 2000; Nucifora et al., 2001). Interestingly, treatment of cells not expressing polyglutamine-expanded Htt with curcumin, a specific inhibitor of p300/CBP HAT activity (Balasubramanyam et al., 2004; Marcu et al., 2006), led to the global hypoacetylation of histone H3 and the nonspecific downregulation of gene expression that did not mimic an HD gene profile (Sadri-Vakili et al., 2007). The conclusions reached from these results implicate a factor, or factors, other than the inhibition of CBP-dependent histone acetylation by mutant Htt protein in conferring the specific pathology that is associated with HD.

The tissue-specific nature of polyglutamine-expansion disorders implies that their associated polyglutamine-expanded proteins affect a discreet set of regulatory factors and specific subsets of genes that are particularly vital for the normal function of neuronal cells and tissues. Interestingly, treatment of transgenic animals with HDAC inhibitors, which increase histone acetylation in a nonspecific manner, can alleviate symptoms and some of their underlying causes in these relatively specific diseases. While moving the field in a positive direction, this only highlights the need for a more thorough understanding of how polyglutamine expansion of proteins cause cellular dysfunction and, more generally, how histone acetylation influences cellular functions on a global level. These findings highlight two important observations involving the potential use of HDAC inhibitors in the treatment of polyglutamine-expansion disorders. First, clear evidence exists indicating that HDAC inhibitors can at least partially prevent cytotoxicity associated with the expression of polyglutamine-expanded proteins in both cell culture and transgenic animal models. Second, while several of these chemotherapeutic agents have shown promise in alleviating the toxic effects of polyglutamine-expanded proteins, no specific mechanism has been elucidated and not all HDAC inhibitors exert the same protective activity. While the ameliorative effects of HDAC inhibitors in the context of polyglutamine-expansion disorders is promising in the laboratory setting, the translation of these findings to accepted and effective clinical applications is still years away. Despite this realization, it appears that recent advances in the understanding of disease mechanisms and possible avenues for clinical treatment of polyglutamine-expansion disorders provide a strong foundation for future findings.

Acknowledgments

The authors would like to thank Paul J. Bonthuis and Madison E. Stellfox for their critical review of this chapter. This work is supported by NIH grant #R01NS049065.

References

Balasubramanian, R., Pray-Grant, M. G., Selleck, W., Grant, P. A., Tan, S., (2002). Role of the Ada2 and Ada3 transcriptional coactivators in histone acetylation. *J. Biol. Chem.* **277**, 7989–7995.

Balasubramanyam, K., Varier, R. A., Altaf, M., Swaminathan, V., Siddappa, N. B., Ranga, U., et al., (2004). Curcumin, a novel p300/CREB-binding protein-specific inhibitor of acetyltransferase, represses the acetylation of histone/nonhistone proteins and histone acetyltransferase-dependent chromatin transcription. *J. Biol. Chem.* **279**, 51163–51171.

Barlev, N. A., Emelyanov, A. V., Castagnino, P., Zegerman, P., Bannister, A. J., Sepulveda, M. A., et al., (2003). A novel human Ada2 homologue functions with Gcn5 or Brg1 to coactivate transcription. *Mol. Cell. Biol.* **23**, 6944–6957.

Benomar, A., Krols, L., Stevanin, G., Cancel, G., LeGuern, E., David, G., et al., (1995). The gene for autosomal dominant cerebellar ataxia with pigmentary macular dystrophy maps to chromosome 3p12-p21.1. *Nat. Genet.* **10**, 84–88.

Benomar, A., Le Guern, E., Durr, A., Ouhabi, H., Stevanin, G., Yahyaoui, M., et al., (1994). Autosomal-dominant cerebellar ataxia with retinal degeneration (ADCA type II) is genetically different from ADCA type I. *Ann. Neurol.* **35**, 439–444.

Berke, S. J., Chai, Y., Marrs, G. L., Wen, H. and Paulson, H. L., (2005). Defining the role of ubiquitin-interacting motifs in the polyglutamine disease protein, ataxin-3. *J. Biol. Chem.* **280**, 32026–32034.

Brownell, J. E. and Allis, C. D., (1995). An activity gel assay detects a single, catalytically active histone acetyltransferase subunit in Tetrahymena macronuclei. *Proc. Natl. Acad. Sci. U.S.A.* **92**, 6364–6368.

Brownell, J. E., Zhou, J., Ranalli, T., Kobayashi, R., Edmondson, D. G., Roth, S. Y., et al., (1996). Tetrahymena histone acetyltransferase A: a homolog to yeast Gcn5p linking histone acetylation to gene activation. *Cell.* **84**, 843–851.

Burnett, B. G. and Pittman, R. N., (2005). The polyglutamine neurodegenerative protein ataxin 3 regulates aggresome formation. *Proc. Natl. Acad. Sci. U.S.A.* **102**, 4330–4335.

Candau, R., Moore, P. A., Wang, L., Barlev, N., Ying, C. Y., Rosen, C. A., et al., (1996). Identification of human proteins functionally conserved with the yeast putative adaptors ADA2 and GCN5. *Mol. Cell. Biol.* **16**, 593–602.

Chen, S., Peng, G. H., Wang, X., Smith, A. C., Grote, S. K., Sopher, B. L., et al., (2004). Interference of Crx-dependent transcription by ataxin-7 involves interaction between the glutamine regions and requires the ataxin-7 carboxy-terminal region for nuclear localization. *Hum. Mol. Genet.* **13**, 53–67.

Ciurciu, A., Komonyi, O., Pankotai, T., Boros, I. M., (2006). The Drosophila histone acetyltransferase Gcn5 and transcriptional adaptor Ada2a are involved in nucleosomal histone H4 acetylation. *Mol. Cell. Biol.* **26**, 9413–9423.

Courey, A. J. Tjian, R., (1988). Analysis of Sp1 in vivo reveals multiple transcriptional domains, including a novel glutamine-rich activation motif. *Cell.* **55**, 887–898.

Custer, S. K., Garden, G. A., Gill, N., Rueb, U., Libby, R. T., Schultz, C., et al., (2006). Bergmann glia expression of polyglutamine-expanded ataxin-7 produces neurodegeneration by impairing glutamate transport. *Nat. Neurosci.* **9**, 1302–1311.

Daniel, J. A., Torok, M. S., Sun, Z. W., Schieltz, D., Allis, C. D., Yates, J. R., 3rd, et al., (2004). Deubiquitination of histone H2B by a yeast acetyltransferase complex regulates transcription. *J. Biol. Chem.* **279**, 1867–1871.

David, G., Abbas, N., Stevanin, G., Durr, A., Yvert, G., Cancel, G., et al., (1997). Cloning of the SCA7 gene reveals a highly unstable CAG repeat expansion. *Nat. Genet.* **17**, 65–70.

David, G., Giunti, P., Abbas, N., Coullin, P., Stevanin, G., Horta, W., et al., (1996). The gene for autosomal dominant cerebellar ataxia type II is located in a 5-cM region in 3p12-p13: genetic and physical mapping of the SCA7 locus. *Am. J. Hum. Genet.* **59**, 1328–1336.

Dhalluin, C., Carlson, J. E., Zeng, L., He, C., Aggarwal, A. K., Zhou, M. M., (1999). Structure and ligand of a histone acetyltransferase bromodomain. *Nature*. **399**, 491–496.

Eberharter, A., Sterner, D. E., Schieltz, D., Hassan, A., Yates, J. R.,3rd, Berger, S. L., et al., (1999). The ADA complex is a distinct histone acetyltransferase complex in Saccharomyces cerevisiae. *Mol. Cell. Biol.* **19**, 6621–6631.

Ellerby, L. M., Hackam, A. S., Propp, S. S., Ellerby, H. M., Rabizadeh, S., Cashman, N. R., et al., (1999). Kennedy's disease: caspase cleavage of the androgen receptor is a crucial event in cytotoxicity. *J. Neurochem.* **72**, 185–195.

Evert, B. O., Araujo, J., Vieira-Saecker, A. M., de Vos, R. A., Harendza, S., Klockgether, T., et al., (2006). Ataxin-3 represses transcription via chromatin binding, interaction with histone deacetylase 3, and histone deacetylation. *J. Neurosci.* **26**, 11474–11486.

Gamper, A. M., Kim, J., Roeder, R. G., (2009). The STAGA subunit ADA2b is an important regulator of human GCN5 catalysis. *Mol. Cell. Biol.* **29**, 266–280.

Garden, G. A., Libby, R. T., Fu, Y. H., Kinoshita, Y., Huang, J., Possin, D. E., et al., (2002). Polyglutamine-expanded ataxin-7 promotes non-cell-autonomous purkinje cell degeneration and displays proteolytic cleavage in ataxic transgenic mice. *J. Neurosci.* **22**, 4897–4905.

Georgakopoulos, T., Thireos, G., (1992). Two distinct yeast transcriptional activators require the function of the GCN5 protein to promote normal levels of transcription. *EMBO J.* **11**, 4145–4152.

Gerber, H. P., Seipel, K., Georgiev, O., Hofferer, M., Hug, M., Rusconi, S., et al., (1994). Transcriptional activation modulated by homopolymeric glutamine and proline stretches. *Science*. **263**, 808–811.

Grant, P. A., Berger, S. L., (1999). Histone acetyltransferase complexes. *Semin. Cell Dev. Biol.* **10**, 169–177.

Grant, P. A., Duggan, L., Cote, J., Roberts, S. M., Brownell, J. E., Candau, R., et al., (1997). Yeast Gcn5 functions in two multisubunit complexes to acetylate nucleosomal histones: characterization of an Ada complex and the SAGA (Spt/Ada) complex. *Genes Dev.* **11**, 1640–1650.

Grant, P. A., Eberharter, A., John, S., Cook, R. G., Turner, B. M., Workman, J. L., (1999). Expanded lysine acetylation specificity of Gcn5 in native complexes. *J. Biol. Chem.* **274**, 5895–5900.

Grant, P. A., Schieltz, D., Pray-Grant, M. G., Steger, D. J., Reese, J. C., Yates, J. R., 3rd, et al., (1998). A subset of TAF(II)s are integral components of the SAGA complex required for nucleosome acetylation and transcriptional stimulation. *Cell.* **94**, 45–53.

Guelman, S., Suganuma, T., Florens, L., Swanson, S. K., Kiesecker, C. L., Kusch, T. et al., (2006). Host cell factor and an uncharacterized SANT domain protein are stable components of ATAC, a novel dAda2A/dGcn5-containing histone acetyltransferase complex in Drosophila. *Mol. Cell. Biol.* **26**, 871–882.

Helmlinger, D., Abou-Sleymane, G., Yvert, G., Rousseau, S., Weber, C., Trottier, Y., et al., (2004). Disease progression despite early loss of polyglutamine protein expression in SCA7 mouse model. *J. Neurosci.* **24**, 1881–1887.

Helmlinger, D., Hardy, S., Abou-Sleymane, G., Eberlin, A., Bowman, A. B., Gansmuller, A., et al., (2006). Glutamine-expanded ataxin-7 alters TFTC/STAGA recruitment and chromatin structure leading to photoreceptor dysfunction. *PLoS Biol.* **4**, e67.

Helmlinger, D., Hardy, S., Sasorith, S., Klein, F., Robert, F., Weber, C., et al., (2004). Ataxin-7 is a subunit of GCN5 histone acetyltransferase-containing complexes. *Hum. Mol. Genet.* **13**, 1257–1265.

Hennig, A. K., Peng, G. H., Chen, S., (2008). Regulation of photoreceptor gene expression by Crx-associated transcription factor network. *Brain Res.* **1192**, 114–133.

Henry, K. W., Wyce, A., Lo, W. S., Duggan, L. J., Emre, N. C., Kao, C. F., et al., (2003). Transcriptional activation via sequential histone H2B ubiquitylation and deubiquitylation, mediated by SAGA-associated Ubp8. *Genes Dev.* **17**, 2648–2663.

Horiuchi, J., Silverman, N., Marcus, G. A., Guarente, L., (1995). ADA3, a putative transcriptional adaptor, consists of two separable domains and interacts with ADA2 and GCN5 in a trimeric complex. *Mol. Cell. Biol.* **15**, 1203–1209.

Ito, T., Chiba, T., Ozawa, R., Yoshida, M., Hattori, M., Sakaki, Y., (2001). A comprehensive two-hybrid analysis to explore the yeast protein interactome. *Proc. Natl. Acad. Sci. U.S.A.* **98**, 4569–4574.

Jacobson, R. H., Ladurner, A. G., King, D. S., Tjian, R., (2000). Structure and function of a human TAFII250 double bromodomain module. *Science.* **288**, 1422–1425.

James, N., Landrieux, E., Collart, M. A., (2007). A SAGA-independent function of SPT3 mediates transcriptional deregulation in a mutant of the Ccr4-not complex in Saccharomyces cerevisiae. *Genetics.* **177**, 123–135.

Kikuchi, H., Takami, Y., Nakayama, T., (2005). GCN5: a supervisor in all-inclusive control of vertebrate cell cycle progression through transcription regulation of various cell cycle-related genes. *Gene.* **347**, 83–97.

Kohler, A., Pascual-Garcia, P., Llopis, A., Zapater, M., Posas, F., Hurt, E., et al., (2006). The mRNA export factor Sus1 is involved in Spt/Ada/Gcn5 acetyltransferase-mediated H2B deubiquitinylation through its interaction with Ubp8 and Sgf11. *Mol. Biol. Cell.* **17**, 4228–4236.

Kohler, A., Schneider, M., Cabal, G. G., Nehrbass, U., Hurt, E., (2008). Yeast Ataxin-7 links histone deubiquitination with gene gating and mRNA export. *Nat. Cell. Biol.* **10**, 707–715.

Kuo, M. H., Brownell, J. E., Sobel, R. E., Ranalli, T. A., Cook, R. G., Edmondson, D. G., et al., (1996). Transcription-linked acetylation by Gcn5p of histones H3 and H4 at specific lysines. *Nature.* **383**, 269–272.

Kusch, T., Guelman, S., Abmayr, S. M., Workman, J. L., (2003). Two Drosophila Ada2 homologues function in different multiprotein complexes. *Mol. Cell. Biol.* **23**, 3305–3319.

La Spada, A. R., Fu, Y. H., Sopher, B. L., Libby, R. T., Wang, X., Li, L. Y., et al., (2001). Polyglutamine-expanded ataxin-7 antagonizes CRX function and induces cone-rod dystrophy in a mouse model of SCA7. *Neuron.* **31**, 913–927.

Lee, K. K., Swanson, S. K., Florens, L., Washburn, M. P., Workman, J. L., (2009). Yeast Sgf73/Ataxin-7 serves to anchor the deubiquitination module into both SAGA and Slik(SALSA) HAT complexes. *Epigenetics Chromatin.* **2**, 2.

Leranth, C., Hamori, J., (1970). "Dark" Purkinje cells of the cerebellar cortex. *Acta Biol. Acad. Sci. Hung.* **21**, 405–419.

Li, F., Macfarlan, T., Pittman, R. N. and Chakravarti, D., (2002). Ataxin-3 is a histone-binding protein with two independent transcriptional corepressor activities. *J. Biol. Chem.* **277**, 45004–45012.

Lindblad, K., Savontaus, M. L., Stevanin, G., Holmberg, M., Digre, K., Zander, C., et al., (1996). An expanded CAG repeat sequence in spinocerebellar ataxia type 7. *Genome Res.* **6**, 965–971.

Luger, K., Mader, A. W., Richmond, R. K., Sargent, D. F., Richmond, T. J., (1997). Crystal structure of the nucleosome core particle at 2.8 A resolution. *Nature.* **389**, 251–260.

Marcu, M. G., Jung, Y. J., Lee, S., Chung, E. J., Lee, M. J., Trepel, J., et al., (2006). Curcumin is an inhibitor of p300 histone acetylatransferase. *Med. Chem.* **2**, 169–174.

Marcus, G. A., Silverman, N., Berger, S. L., Horiuchi, J. Guarente, L., (1994). Functional similarity and physical association between GCN5 and ADA2: putative transcriptional adaptors. *EMBO J.* **13**, 4807–4815.

Martin-Aparicio, E., Yamamoto, A., Hernandez, F., Hen, R., Avila, J., Lucas, J. J., (2001). Proteasomal-dependent aggregate reversal and absence of cell death in a conditional mouse model of Huntington's disease. *J. Neurosci.* **21**, 8772–8781.

Martinez, E., Kundu, T. K., Fu, J., Roeder, R. G., (1998). A human SPT3-TAFII31-GCN5-L acetylase complex distinct from transcription factor IID. *J. Biol. Chem.* **273**, 23781–23785.

Martinez, E., Palhan, V. B., Tjernberg, A., Lymar, E. S., Gamper, A. M., Kundu, T. K., et al., (2001). Human STAGA complex is a chromatin-acetylating transcription coactivator that interacts with pre-mRNA splicing and DNA damage-binding factors in vivo. *Mol. Cell. Biol.* **21**, 6782–6795.

Maurice, T., Duclot, F., Meunier, J., Naert, G., Givalois, L., Meffre, J., et al., (2008). Altered memory capacities and response to stress in p300/CBP-associated factor (PCAF) histone acetylase knockout mice. *Neuropsychopharmacology.* **33**, 1584–1602.

Mayr, B. M., Guzman, E., Montminy, M., (2005). Glutamine rich and basic region/leucine zipper (bZIP) domains stabilize cAMP-response element-binding protein (CREB) binding to chromatin. *J. Biol. Chem.* **280**, 15103–15110.

McCampbell, A., Taye, A. A., Whitty, L., Penney, E., Steffan, J. S., Fischbeck, K. H., (2001). Histone deacetylase inhibitors reduce polyglutamine toxicity. *Proc. Natl. Acad. Sci. U.S.A.* **98**, 15179–15184.

McMahon, S. J., Pray-Grant, M. G., Schieltz, D., Yates, J. R., 3rd, Grant, P. A., (2005). Polyglutamine-expanded spinocerebellar ataxia-7 protein disrupts normal

SAGA and SLIK histone acetyltransferase activity. *Proc. Natl. Acad. Sci. U.S.A.* **102**, 8478–8482.

Minamiyama, M., Katsuno, M., Adachi, H., Waza, M., Sang, C., Kobayashi, Y., et al., (2004). Sodium butyrate ameliorates phenotypic expression in a transgenic mouse model of spinal and bulbar muscular atrophy. *Hum. Mol. Genet.* **13**, 1183–1192.

Mitchell, P. J., Tjian, R., (1989). Transcriptional regulation in mammalian cells by sequence-specific DNA binding proteins. *Science*. **245**, 371–378.

Muratoglu, S., Georgieva, S., Papai, G., Scheer, E., Enunlu, I., Komonyi, O., et al., (2003). Two different Drosophila ADA2 homologues are present in distinct GCN5 histone acetyltransferase-containing complexes. *Mol. Cell. Biol.* **23**, 306–321.

Nagai, Y., Popiel, H. A., (2008). Conformational changes and aggregation of expanded polyglutamine proteins as therapeutic targets of the polyglutamine diseases: exposed beta-sheet hypothesis. *Curr. Pharm. Des.* **14**, 3267–3279.

Nagy, Z., Tora, L., (2007). Distinct GCN5/PCAF-containing complexes function as co-activators and are involved in transcription factor and global histone acetylation. *Oncogene*. **26**, 5341–5357.

Nucifora, F. C.,Jr, Sasaki, M., Peters, M. F., Huang, H., Cooper, J. K., Yamada, M., et al., (2001). Interference by huntingtin and atrophin-1 with cbp-mediated transcription leading to cellular toxicity. *Science*. **291**, 2423–2428.

Ogryzko, V. V., Kotani, T., Zhang, X., Schiltz, R. L., Howard, T., Yang, X. J., et al., (1998). Histone-like TAFs within the PCAF histone acetylase complex. *Cell*. **94**, 35–44.

Orr, H. T., Zoghbi, H. Y., (2007). Trinucleotide repeat disorders. *Annu. Rev. Neurosci.* **30**, 575–621.

Osley, M. A., (2006). Regulation of histone H2A and H2B ubiquitylation. *Brief Funct. Genomic Proteomic*. **5**, 179–189.

Palhan, V. B., Chen, S., Peng, G. H., Tjernberg, A., Gamper, A. M., Fan, Y., et al., (2005). Polyglutamine-expanded ataxin-7 inhibits STAGA histone acetyltransferase activity to produce retinal degeneration. *Proc. Natl. Acad. Sci. U.S.A.* **102**, 8472–8477.

Pray-Grant, M. G., Schieltz, D., McMahon, S. J., Wood, J. M., Kennedy, E. L., Cook, R. G., et al., (2002). The novel SLIK histone acetyltransferase complex functions in the yeast retrograde response pathway. *Mol. Cell. Biol.* **22**, 8774–8786.

Qi, D., Larsson, J., Mannervik, M., (2004). Drosophila Ada2b is required for viability and normal histone H3 acetylation. *Mol. Cell. Biol.* **24**, 8080–8089.

Robzyk, K., Recht, J., Osley, M. A., (2000). Rad6-dependent ubiquitination of histone H2B in yeast. *Science*. **287**, 501–504.

Roth, S. Y., Denu, J. M., Allis, C. D., (2001). Histone acetyltransferases. *Annu. Rev. Biochem.* **70**, 81–120.

Sadri-Vakili, G., Bouzou, B., Benn, C. L., Kim, M. O., Chawla, P., Overland, R. P., et al., (2007). Histones associated with downregulated genes are hypo-acetylated in Huntington's disease models. *Hum. Mol. Genet.* **16**, 1293–1306.

Sanders, S. L., Garbett, K. A., Weil, P. A., (2002). Molecular characterization of Saccharomyces cerevisiae TFIID. *Mol. Cell. Biol.* **22**, 6000–6013.

Shilatifard, A., (2006). Chromatin modifications by methylation and ubiquitination: implications in the regulation of gene expression. *Annu. Rev. Biochem.* **75**, 243–269.

Shogren-Knaak, M., Ishii, H., Sun, J. M., Pazin, M. J., Davie, J. R., Peterson, C. L., (2006). Histone H4-K16 acetylation controls chromatin structure and protein interactions. *Science*. **311**, 844–847.

Shukla, A., Bajwa, P., Bhaumik, S. R., (2006). SAGA-associated Sgf73p facilitates formation of the preinitiation complex assembly at the promoters either in a HAT-dependent or independent manner in vivo. *Nucleic Acids Res.* **34**, 6225–6232.

Smith, E. R., Belote, J. M., Schiltz, R. L., Yang, X. J., Moore, P. A., Berger, S. L., et al., (1998). Cloning of Drosophila GCN5: conserved features among metazoan GCN5 family members. *Nucleic Acids Res.* **26**, 2948–2954.

Smith, K. T., Workman, J. L., (2009). Histone deacetylase inhibitors: anticancer compounds. *Int. J. Biochem. Cell Biol.* **41**, 21–25.

Steffan, J. S., Kazantsev, A., Spasic-Boskovic, O., Greenwald, M., Zhu, Y. Z., Gohler, H., et al., (2000). The Huntington's disease protein interacts with p53 and CREB-binding protein and represses transcription. *Proc. Natl. Acad. Sci. U.S.A.* **97**, 6763–6768.

Sterner, D. E., Belotserkovskaya, R., Berger, S. L., (2002). SALSA, a variant of yeast SAGA, contains truncated Spt7, which correlates with activated transcription. *Proc. Natl. Acad. Sci. U.S.A.* **99**, 11622–11627.

Strahl, B. D., Allis, C. D., (2000). The language of covalent histone modifications. *Nature*. **403**, 41–45.

Strahlendorf, J. C., Brandon, T., Miles, R., Strahlendorf, H. K., (1998). AMPA receptor-mediated alterations of intracellular calcium homeostasis in rat cerebellar Purkinje cells in vitro: correlates to dark cell degeneration. *Neurochem. Res.* **23**, 1355–1362.

Strahlendorf, J. C., Strahlendorf, H. K., (1999). Enduring changes in Purkinje cell electrophysiology following transient exposure to AMPA: correlates to dark cell degeneration. *Neurosci. Res.* **33**, 155–162.

Taverna, S. D., Li, H., Ruthenburg, A. J., Allis, C. D., Patel, D. J. (2007). How chromatin-binding modules interpret histone modifications: lessons from professional pocket pickers. *Nat. Struct. Mol. Biol.* **14**, 1025–1040.

Titz, B., Thomas, S., Rajagopala, S. V., Chiba, T., Ito, T., Uetz, P., (2006). Transcriptional activators in yeast. *Nucleic Acids Res.* **34**, 955–967.

Torok, M. S., Grant, P. A. (2004). Histone acetyltransferase proteins contribute to transcriptional processes at multiple levels. Adv. Protein Chem. **67**, 181–199.

Uetz, P., Giot, L., Cagney, G., Mansfield, T. A., Judson, R. S., Knight, J. R., et al., (2000). A comprehensive analysis of protein-protein interactions in Saccharomyces cerevisiae. *Nature*. **403**, 623–627.

Wang, Y. L., Faiola, F., Xu, M., Pan, S., Martinez, E. (2008). Human ATAC Is a GCN5/PCAF-containing acetylase complex with a novel NC2-like histone fold module that interacts with the TATA-binding protein. J. Biol. Chem. **283**, 33808–33815.

Weake, V. M., Lee, K. K., Guelman, S., Lin, C. H., Seidel, C., Abmayr, S. M., et al., (2008). SAGA-mediated H2B deubiquitination controls the development of neuronal connectivity in the Drosophila visual system. *EMBO J.* **27**, 394–405.

Weake, V. M., Workman, J. L., (2008). Histone ubiquitination: triggering gene activity. *Mol. Cell.* **29**, 653–663.

Wellington, C. L., Ellerby, L. M., Hackam, A. S., Margolis, R. L., Trifiro, M. A., Singaraja, R., et al., (1998). Caspase cleavage of gene products associated with triplet expansion disorders generates truncated fragments containing the polyglutamine tract. *J. Biol. Chem.* **273**, 9158–9167.

Wyce, A., Xiao, T., Whelan, K. A., Kosman, C., Walter, W., Eick, D., et al., (2007). H2B ubiquitylation acts as a barrier to Ctk1 nucleosomal recruitment prior to removal by Ubp8 within a SAGA-related complex. *Mol. Cell.* **27**, 275–288.

Xu, W., Edmondson, D. G., Evrard, Y. A., Wakamiya, M., Behringer, R. R., Roth, S. Y., (2000). Loss of Gcn5l2 leads to increased apoptosis and mesodermal defects during mouse development. *Nat. Genet.* **26**, 229–232.

Xu, W., Edmondson, D. G., Roth, S. Y., (1998). Mammalian GCN5 and P/CAF acetyltransferases have homologous amino-terminal domains important for recognition of nucleosomal substrates. *Mol. Cell. Biol.* **18**, 5659–5669.

Yamauchi, T., Yamauchi, J., Kuwata, T., Tamura, T., Yamashita, T., Bae, N., et al., (2000). Distinct but overlapping roles of histone acetylase PCAF and of the closely related PCAF-B/GCN5 in mouse embryogenesis. *Proc. Natl. Acad. Sci. U.S.A.* **97**, 11303–11306.

Yang, X. J., Ogryzko, V. V., Nishikawa, J., Howard, B. H., Nakatani, Y., (1996). A p300/CBP-associated factor that competes with the adenoviral oncoprotein E1A. *Nature.* **382**, 319–324.

Ying, M., Xu, R., Wu, X., Zhu, H., Zhuang, Y., Han, M., et al., (2006). Sodium butyrate ameliorates histone hypoacetylation and neurodegenerative phenotypes in a mouse model for DRPLA. *J. Biol. Chem.* **281**, 12580–12586.

Yoo, S. Y., Pennesi, M. E., Weeber, E. J., Xu, B., Atkinson, R., Chen, S., et al., (2003). SCA7 knockin mice model human SCA7 and reveal gradual accumulation of mutant ataxin-7 in neurons and abnormalities in short-term plasticity. *Neuron.* **37**, 383–401.

Young, J. E., Gouw, L., Propp, S., Sopher, B. L., Taylor, J., Lin, A., et al., (2007). Proteolytic cleavage of ataxin-7 by caspase-7 modulates cellular toxicity and transcriptional dysregulation. *J. Biol. Chem.* **282**, 30150–30160.

Yvert, G., Lindenberg, K. S., Devys, D., Helmlinger, D., Landwehrmeyer, G. B., Mandel, J. L., (2001). SCA7 mouse models show selective stabilization of mutant ataxin-7 and similar cellular responses in different neuronal cell types. *Hum. Mol. Genet.* **10**, 1679–1692.

Yvert, G., Lindenberg, K. S., Picaud, S., Landwehrmeyer, G. B., Sahel, J. A., Mandel, J. L., (2000). Expanded polyglutamines induce neurodegeneration and transneuronal alterations in cerebellum and retina of SCA7 transgenic mice. *Hum. Mol. Genet.* **9**, 2491–2506.

Zhao, Y., Lang, G., Ito, S., Bonnet, J., Metzger, E., Sawatsubashi, S., et al., (2008). A TFTC/STAGA module mediates histone H2A and H2B deubiquitination, coactivates nuclear receptors, and counteracts heterochromatin silencing. *Mol. Cell.* **29**, 92–101.

AUTHOR INDEX

A

B

C

D

E

F

G

L

M

S

T

SUBJECT INDEX

Note: The letters '*f*' and '*t*' following the locators refer to figures and tables respectively.

A

B

C

D

E

F

G

H

Q

R

S

T

U

V

W

X

Y

(A)

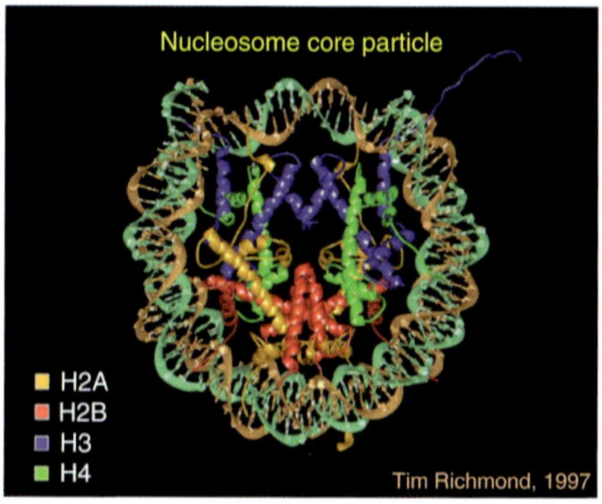

(B)

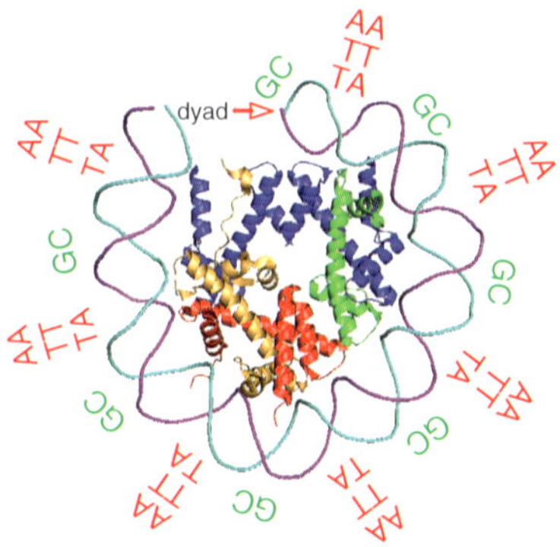

Denis Tolkunov and Alexandre V. Morozov, Chapter 1, Fig 1. (A) Crystal structure of the nucleosome core particle (Luger et al., 1997) (courtesy of Song Tan, Penn State University). The 147-bp long DNA is wrapped around the histone octamer in ~1.65 turns of a left-handed superhelix. The histone octamer consists of two copies each of histones H2A (yellow), H2B (red), H3 (blue), and H4 (green). (B) Bending of nucleosomal DNA is mediated by specific dinucleotides located at positions where DNA minor or major groove faces the histone octamer (reproduced with permission from Segal et al., 2006). Relative frequencies of A/T-rich dinucleotides tend to increase at positions where the minor groove faces the surface of the histone octamer, whereas relative frequencies of G/C-rich dinucleotides tend to increase where the minor groove faces away from and the major groove faces toward the histone octamer.

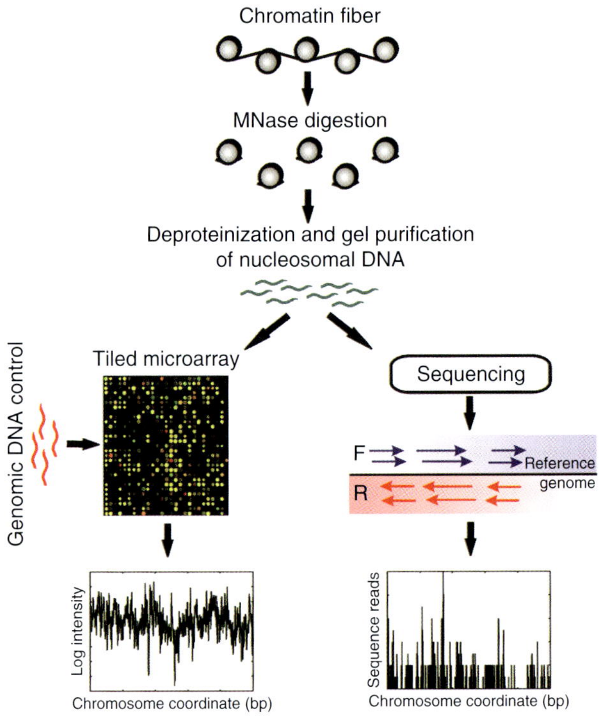

DENIS TOLKUNOV AND ALEXANDRE V. MOROZOV, CHAPTER 1, FIG. 2. Schematic representation of high-throughput nucleosome-positioning experiments. Chromatin is subjected to MNase digestion, the resulting DNA is purified, and mononucleosomal DNA is isolated on a gel. Optionally, histones are cross-linked on DNA prior to the MNase digestion and immunoprecipitated, after which the cross-linking is reversed. In a chip-based experiment (left panel), mononucleosomal DNA is hybridized to a microarray together with a control sample prepared by digesting genomic nucleosome-free DNA with MNase. The log ratio of nucleosomal DNA intensity to genomic DNA intensity is plotted as a function of each probe's starting position on the genome. Higher-than-average log-intensity values correspond to nucleosome-covered regions, while lower-than-average log-intensity values correspond to nucleosome-depleted regions. Microarray probes are tiled across the entire genome or genomic regions of interest. In a high-throughput parallel sequencing experiment, mononucleosomal (and in some cases control) DNA is sequenced directly (right panel). High-throughput sequencing yields large collections of reads typically shorter than the DNA lengths in the input sample. The reads are mapped onto both strands of the reference genome (often with several mismatches allowed) and combined into a single sequence-read profile by assuming that each nucleosome core particle has a fixed length of 147 bp. With this assumption, sequence reads from the reverse strand are remapped onto the forward strand by subtracting 147 bp from their start coordinates. The resulting sequence-read profile contains information about the number of reads assigned to every genomic position.

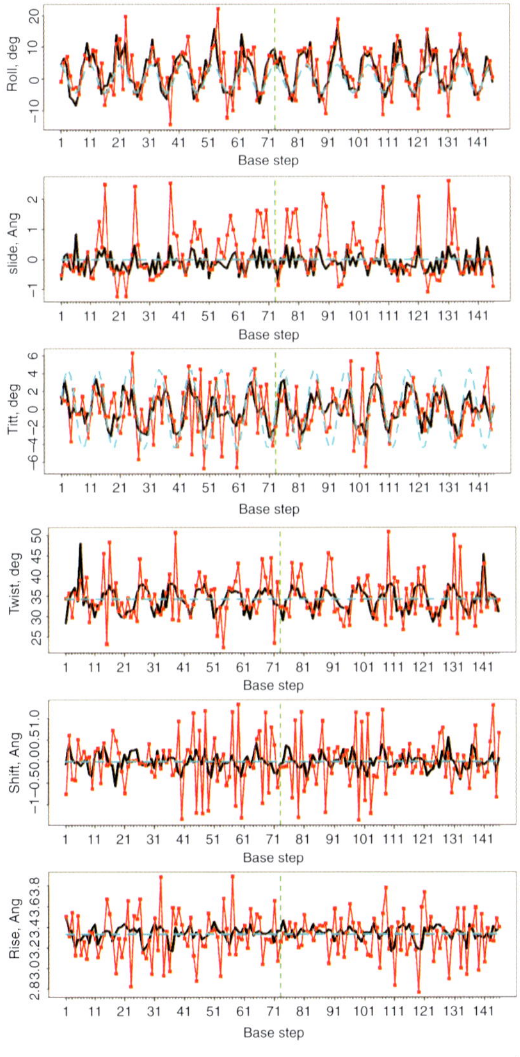

Denis Tolkunov and Alexandre V. Morozov, Chapter 1, Fig 7. Dinucleotide geometries from the crystal structure of the nucleosome core particle (solid red lines with squares, PDB code 1kx5) (Richmond and Davey, 2003): from the minimum-energy structure obtained by relaxing the 147-bp DNA segment with the nucleotide sequence from 1kx5 around the ideal superhelix as described in Section III.A (solid black lines) (Morozov et al., 2009); and from the ideal superhelix (dashed cyan lines). Six dinucleotide degrees of freedom are shown: roll, slide, tilt, twist, shift, and rise. The twofold nucleosome symmetry axis is shown as a dashed vertical line. Note that roll is negative when DNA bends into the minor groove (which faces the histone octamer) and positive when DNA bends into the major groove (with the minor groove facing away from the histone octamer).

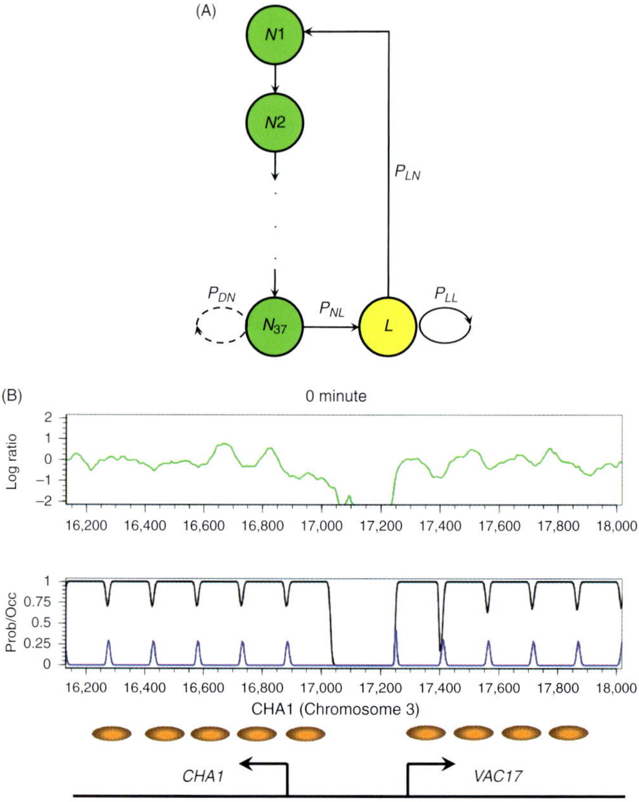

Denis Tolkunov and Alexandre V. Morozov, Chapter 1, Fig 11. (A) The transition matrix of the Hidden Markov Model (HMM) used for predicting nucleosome probabilities and occupancies from microarray log-intensity profiles. Nodes N_1 through N_{37} represent consecutive nucleosomal states (green circles), while L is the linker state (yellow circle). Once a new nucleosome is started 36 subsequent nodes are placed with a probability 1. From N_{37} a transition is made to the linker state with a probability P_{NL}, or else the nucleosome is extended indefinitely to create "delocalized" nucleosomal states (Yuan et al., 2005; Lee et al., 2007). The linker state can also be extended indefinitely with a probability P_{LL}, or else the transition is made back to the first nucleosome state N_1. Note that $P_{NL}+P_{DN}=1$, $P_{LL}+P_{LN}=1$. (B) Nucleosome positions prior to glucose addition at the *CHA1* promoter predicted by the HMM. Top: Log ratio of nucleosomal DNA to genomic DNA as determined by separate hybridizations to Affymetrix tiling arrays is plotted as a function of genomic position. Increasing values represent increasing protection from MNase digestion. Middle: Nucleosome positions predicted by the HMM (Zawadzki et al., 2009). Black trace represents predicted nucleosome occupancy, from unoccupied (0) to fully occupied (1). Blue trace represents the probability of starting a nucleosome at a given base pair. Bottom: Previously mapped *in vivo* nucleosome positions (Moreira and Holmberg, 1998) are shown as dark orange ovals.

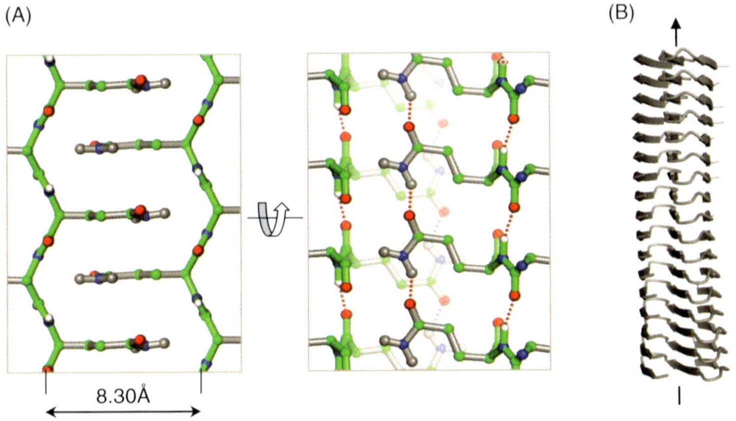

JULIEN JORDA AND ANDREY V. KAJAVA, CHAPTER 2, FIG 5. (A) A model of β-crystallite of poly-Q with an orthogonal unit cell with $a = 9.5$ Å, $b = 16.60$ Å, and $c = 6.95$ Å (chain axis) (Sikorski and Atkins, 2005). On the left, a view along the H-bond directions shows interdigitation of Q side chains. On the right, a view along the β-strands illustrates how the H-bonding of Q side chains creates ladders inside the β-structure. Sticks of the main chains are in green and of the side chains are in gray. (B) A schematic diagram of a cross-β amyloid fibril formed by twisted β-sheets. Beta-strands are shown as arrows. The twist can delimit lateral dimensions and favour axial growth of the homopeptide aggregates. When considered locally, side chain packings (e.g., the H-bonded Q-ladders) within the crystallite (A) and fibril-like assemblies (B) can be similar.

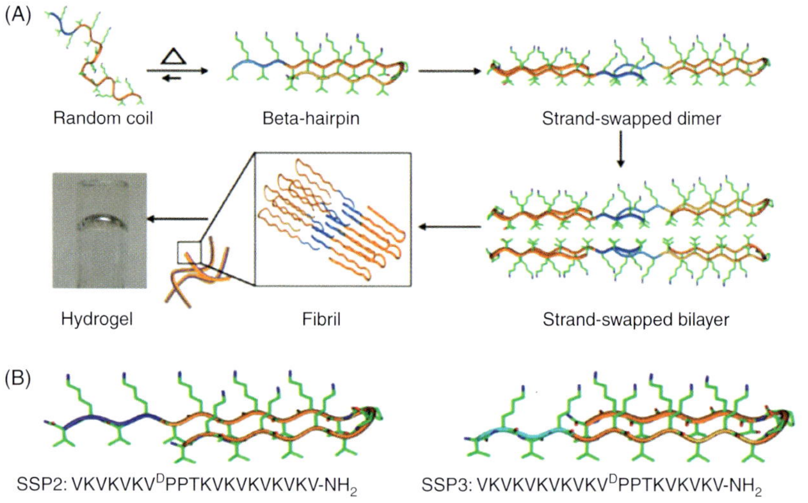

Anna Mitraki, Chapter 3, Fig 8. "Domain swapping" in materials design. (A) A designed peptide has a random coil structure, folds into a beta-hairpin upon heating, and subsequently self-assembles and forms hydrogels. (B) Schematic drawing of the designed peptides. The blue part depicts an "exchangeable" sequence that is capable for "strand swapping" located at the N-terminus of the peptide. The cyan part depicts an "exchangeable" sequence that is capable for "strand swapping" located at the C-terminus of the peptide. The "nonexchangeable" domain is depicted in orange. The sequences of the two peptides in one-letter code are shown at the bottom. From Nagarkar et al. (2010), reprinted with permission from Wiley.

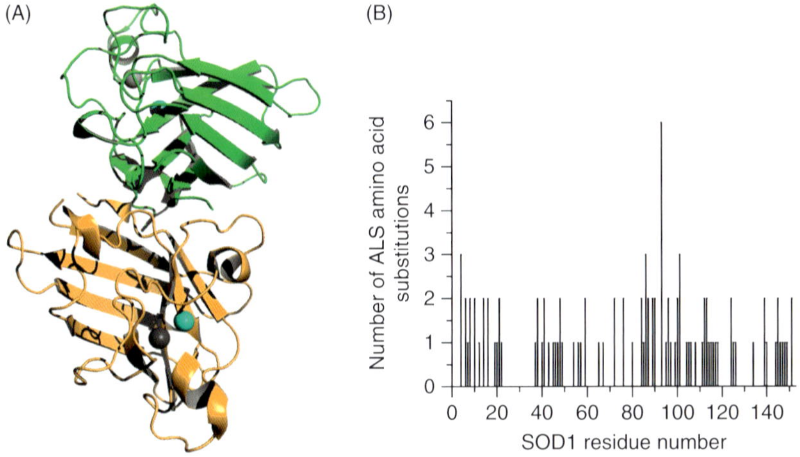

BRYAN F. SHAW ET AL., CHAPTER 4, FIG 1. (A) Ribbon diagram of human Cu, Zn superoxide dismutase 1 (SOD1). Different subunits of the homodimeric protein are colored orange and green. Each subunit binds one copper (cyan sphere) and one zinc ion (charcoal sphere). The fold of each subunit consists of an eight-stranded, antiparallel, Greek-key β-barrel (because the β-barrel of SOD1 is not completely closed, the fold of SOD1 is better described as a "β-taco"). PDB Code: 2V0A. (B) Most mutations in SOD1 that cause ALS (>90%) result in single amino acid substitutions. This frequency plot shows where each known ALS-linked amino acid substitution occurs in the 153-residue SOD1 polypeptide. The substitutions occur throughout the polypeptide chain, but glycine 93 appears to represent a "hot spot"; six different substitutions occur at Gly 93.

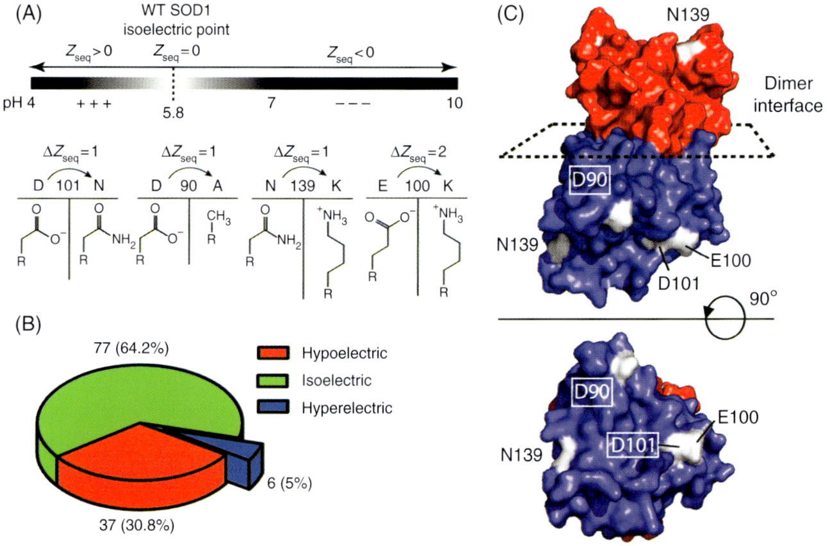

BRYAN F. SHAW ET AL., CHAPTER 4, FIG 2. (A) Wild-type (WT) human superoxide dismutase 1 (SOD1) is predicted to have a net negative charge at pH > 5.8. The calculated net charge (Z_{seq}) of the SOD1 monomer at pH 7.4 is −6 (for the homodimer $Z_{seq} = -12$). The *cryptic* mutants: D90A, D101N, N139K, and E100K do not significantly alter the structure, conformational stability, metal binding, or folding properties of the SOD1 protein. Each of these mutations does, however, decrease the formal net negative charge of SOD1 by 1 unit or 2 units at pH 7.4. (B) An analysis of 120 ALS mutations to SOD1 (available online at www.alsa.org) shows that 114 mutations (95%) are either isoelectric (e.g., do not change Z_{seq} at physiological pH) or hypoelectric (e.g., decrease Z_{seq}); only 5% are hyperelectric or increase Z_{seq} (1 of these 6 mutations, V7E, also severely destabilizes the folded SOD1 protein (Rodriguez et al., 2005)). (C) Surface rendering of the X-ray crystal structure of WT-SOD1 (PDB code: 2V0A). The D101N, D90A, N139K, and E100K substitutions occur at the surface of SOD1 and away from the dimer interface.

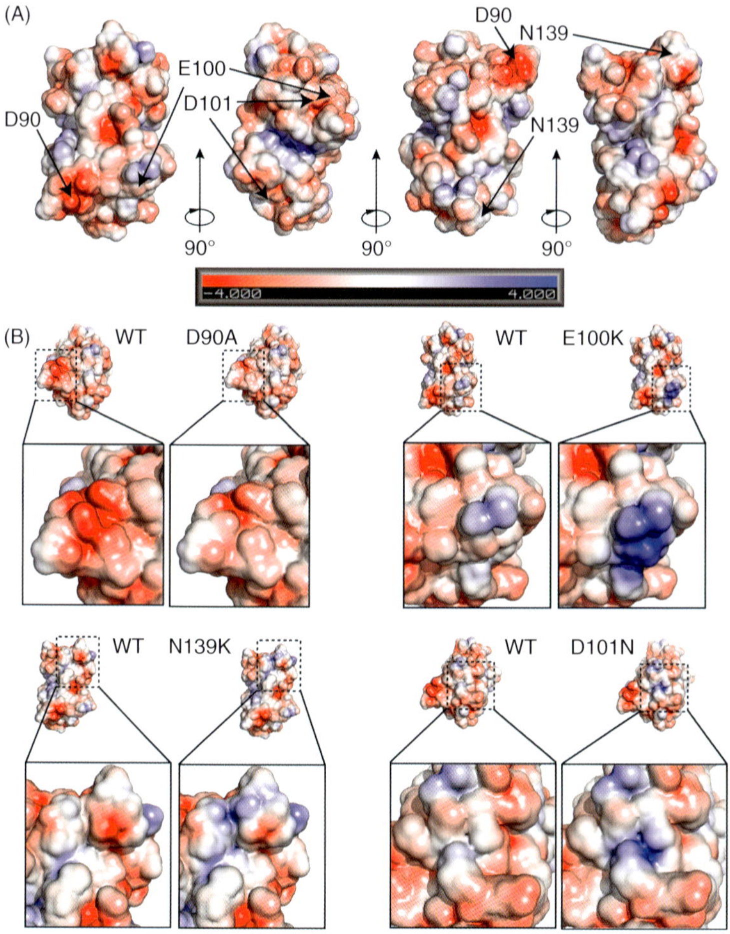

BRYAN F. SHAW ET AL., CHAPTER 4, FIG 3. (A) A nonlinear Poisson–Boltzmann equation was solved (using the APMS software package) to calculate the electrostatic potential distribution of the solvent accessible surface of wild-type human SOD1, based upon the X-ray crystal structure (PDB code: 2V0A); blue indicates positive electrostatic surface potential; red indicates negative electrostatic surface potential. (B) Structures of the 4 cryptic mutants (D90A, E100K, D101N, and N139K) were modeled by substituting each amino acid with wild-type SOD1 (e.g., the X-ray crystal structure of wild-type SOD1) using the MOE software package. We calculated the electrostatic potential distributions using the same method that we applied to the structure of wild-type SOD1. The unit of the color scale in both (A) and (B) is eV. These four cryptic substitutions do not alter the electrostatic potential of a similar region on the surface of SOD1 but affect different areas.

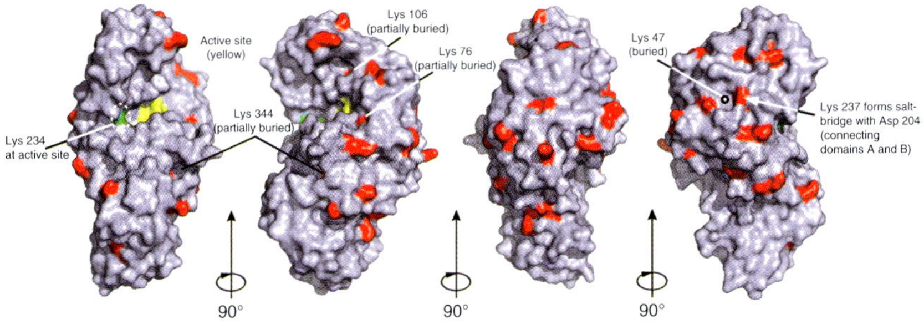

Bryan F. Shaw et al., Chapter 4, Fig 5. Surface rendering of the X-ray crystal structure of the thermostable variant of α-amylase from *Bacillus licheniformis* (N190F/Q264S/N265Y BLA). Lysine residues are colored in red (with the exception of the strictly conserved lysine 234, at the active site, which is colored green). The active site region is colored yellow. The majority of lysine residues are not located on the same face of the protein as the active site, and nearly all 28 lysine residues are solvent exposed. Lysine 76, 106, and 344 are partially buried and Lys 47 is completely buried and not visible with a surface rendering. PDB code:1BLI.

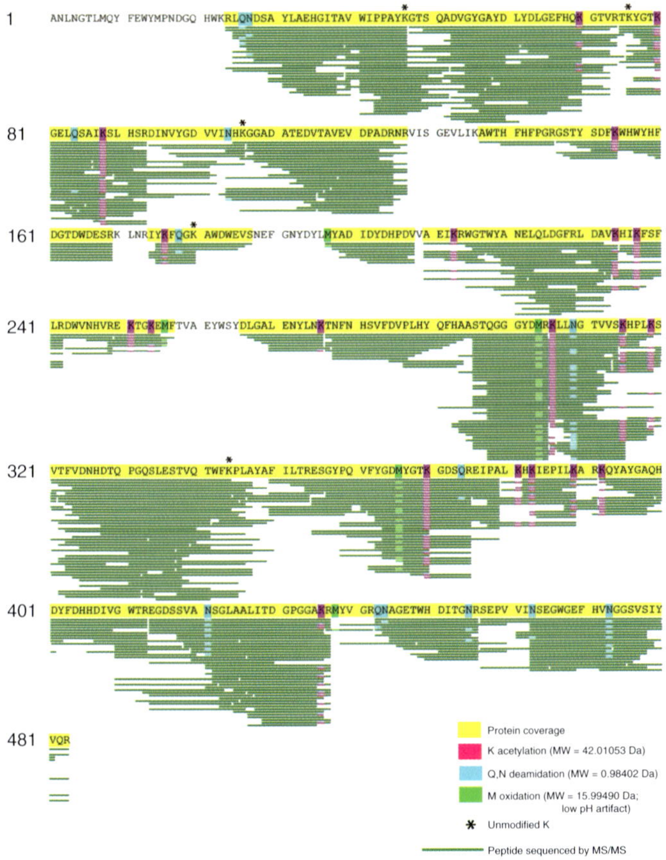

BRYAN F. SHAW ET AL., CHAPTER 4, FIG 9. Identification of acetylated lysine residues in a sample of "BLA-Ac(~17)" with a bottom-up proteomic approach (LC/ESI-MS/MS). Eighteen different lysine residues were found to have a 42-Da modification (e.g., Lys 70, 80, 88, 154, 176, 213, 234, 237, 251, 306, 315, 319, 370, 381, 383, 389, 392, and 436). Mass spectrometric data regarding the acetylation of two lysine residues (e.g., Lys 254 and Lys 276) were ambiguous. Five lysine residues were found to have no modifications (e.g., Lys 47, 76, 106, 180, and 344; denoted with an asterisk (*)). Three lysine residues (e.g., Lys 23, 136, and 170) and the N-terminal alanine 1 were not contained in the covered sequence and the acetylation of these residues can not be verified. BLA-Ac(~17) was proteolyzed with trypsin and elastase and the digest was analyzed with LC/ESI-MS/MS. Approximately 89% of the amino acid sequence of BLA was accounted for. Four methionine residues (e.g., Met 197, 256, 366, and 438) were oxidized (methionine oxidation is usually observed in bottom-up proteomic analyses). Adapted from Shaw and Whitesides (Shaw et al., 2008).